Questions on the Posterior Analytics
(Second Redaction)

Questions on the Posterior Analytics
(*Second Redaction*)

Simon of Faversham

*Critical edition by Iacopo Costa, Gustavo Fernández Walker,
John Longeway, and Ana María Mora-Márquez*

*English Translation by John Longeway
and Matthew Wennemann*

https://www.openbookpublishers.com

©2025 Iacopo Costa, Gustavo Fernández Walker, John Longeway, and Ana María Mora-Márquez (eds), John Longeway and Matthew Wennemann (trans.)

This work is licensed under an Attribution-NonCommercial 4.0 International (CC BY-NC 4.0). This license allows you to share, copy, distribute and transmit the text; to adapt the text for non-commercial purposes of the text providing attribution is made to the authors (but not in any way that suggests that they endorse you or your use of the work). Attribution should include the following information:

Iacopo Costa, Gustavo Fernández Walker, John Longeway, and Ana María Mora-Márquez (eds), John Longeway and Matthew Wennemann (trans.), *Questions on the Posterior Analytics (Second Redaction)*. *Simon of Faversham*. Cambridge, UK: Open Book Publishers, 2025, https://doi.org/10.11647/OBP.0468

Further details about CC BY-NC licenses are available at
https://creativecommons.org/licenses/by-nc/4.0/

All external links were active at the time of publication unless otherwise stated and have been archived via the Internet Archive Wayback Machine at
https://archive.org/web

Digital material and resources associated with this volume are available at
https://doi.org/10.11647/OBP.0468#resources

The Medieval Text Consortium Series: Volume 2
ISSN Print: 2754-0634 | ISSN Digital: 2754-0642

ISBN Paperback: 978-1-80511-602-8 | ISBN Hardback: 978-1-80511-603-5 | ISBN Digital (PDF): 978-1-80511-604-2

DOI: 10.11647/OBP.0468

Cover image: Photo by Sean Babbs, 2020, University of Colorado Boulder Libraries Instruction and Outreach | Collections of Distinction Special Collections, Archives, Government Information, and Maps, CC-BY.

Cover design: Jeevanjot Kaur Nagpal

Published by Open Book Publishers in collaboration with Benson Center Press.

EDITORIAL BOARD

Robert Pasnau, *University of Colorado*
Magdalena Bieniak, *University of Warsaw*
Monica Brinzei, *CNRS Paris*
Russell Friedman, *KU Leuven*
Guy Guldentops, *University of Cologne*
Peter Hartman, *Loyola University Chicago*
Peter King, *University of Toronto*
Christopher Martin, *University of Auckland*
Giorgio Pini, *Fordham University*
Cecilia Trifogli, *University of Oxford*

Typesetting
Jan Maliszewski, *University of Warsaw*

Table of Contents

Introduction ix
Abbreviations 1

Simoni Faversham
Questiones super librum Posteriorum

Liber Primus 2
Liber Secundus 192

Bibliography 303

Simon of Faversham,
Questions on the Posterior Analytics

Book 1   3
Book 2   193

A plain text version of the Latin treatise is available
from Open Book Publishers.

Introduction[1]

Simon of Faversham was a late thirteenth-century arts master whose commentaries survive on many of the central Aristotelian works of logic and natural philosophy. Here we offer an edition and English translation of the second of Simon of Faversham's two sets of questions on Aristotle's *Posterior Analytics*. It is likely that this set of questions comes from Simon's teaching at Paris at some point during the end of the 1270s or the beginning of the 1280s.[2]

This work is founded on the edition and translation that John Longeway made in his 1977 doctoral dissertation at Cornell University.

1. Life[3]

Simon came from Faversham, a river port and market town in Kent.[4] He became an acolyte in 1283, and so he was born perhaps about 1260.[5] After receiving a Master of Arts from Oxford, he decided to become a university

---

[1] This work was partly done with the generous financial support of the Knut and Alice Wallenberg Foundation. We are thankful to John Longeway and Bob Pasnau for the trust they put in us for the revision of Longeway's work, and to Bob Pasnau and Cecilia Trifogli for their invaluable comments and suggestions on earlier versions of this edition.

[2] See Simon of Faversham, *Quaestiones super libro Elenchorum*, ed. Ebbesen et al., pp. 4–5.

[3] This account is but a restructuring of John Longeway's account of Simon's life in the introduction to his editions of his commentaries on *Posterior Analytics*. The rightful author of this part is, then, John Longeway.

[4] Faversham lies about ten miles west of Canterbury on Faversham Creek, which is navigable up to the town for vessels under two hundred tons. It enjoys an extensive shipping trade, as well as an oyster fishery. Founded in Roman times, the town was a port of some importance throughout the Middle Ages. A *witan*, that is, a national council, was held there under Athelstan in 930. The town was assessed as a royal demesne from 1086 on. A Cluniac abbey (later Benedictine) was founded there by King Stephen and Queen Matilda in 1147. The Church of Mary, Mother of Charity, an old church even when Simon was alive, is still there, and has been restored.

[5] *Registrum Epistolarum Fratris Johannis Peckham*, ed. Martin, vol. 3, p. 1033. 'Simon Marchaunt de Faversham' is listed among those ordained by the bishop as acolytes in St. Mary's Church, Faversham, on 25 September 1283. This is also reported in *Register of John Pecham*, ed. Davis et al., vol. 2, p. 203. Wolf, *Die Intellektslehre*, pp. 20, 25 ff., is the first scholar to have noted and argued from these references.

teacher. This necessarily involved a clerical career, and Simon was ordained a sub-deacon at Croydon, near London, on 24 September 1289. On 23 September 1290, Archbishop Peckham ordained him deacon in Bocking,[6] and bestowed probably his first ecclesiastical preferment, making him rector of Preston, a village outside Faversham, on the following day.[7] The rectorship provided an income without the duties of residence or cure of souls. It was customary to provide such sinecures for the support of scholars in the Middle Ages, and Simon received several more in the course of his career.[8]

Although Simon had given up any expectation of marriage in becoming a sub-deacon, he seems never to have become a priest. This was not unusual for those not in a religious order, who had little love for the extra duties involved in the priesthood. He also avoided becoming a Doctor of Theology until very late in his career. Probably he took the degree, having long ago fulfilled its requirements, only when it became clear that he might gain the chancellorship at Oxford, for which that degree was a prerequisite.[9] The Arts faculty had begun as a preparatory school through which one must first pass to begin study for a higher degree, but by the late thirteenth century the originally limited body of knowledge that had once constituted the liberal arts had expanded into something worth a lifetime's attention. Thus, men like Simon, whose interests lay in philosophy and logic rather than theology, law, or medicine, might remain Masters of Arts indefinitely, never passing on to the higher faculties.[10] The practice was still somewhat unusual in Simon's day but became quite common in the fourteenth century.

---

[6] *Registrum Epistolarum Fratris Johannis Peckham*, ed. Martin, vol. 3, pp. 1051, 1053. *Register of John Pecham*, ed. Davis et al., vol. 2, pp. 27, 28. He is described as Master Simon in the note referring to his ordination.

[7] *Registrum Epistolarum Fratris Johannis Peckham*, ed. Martin, vol. 3, p. 1011. *Register of John Pecham*, ed. Davis et al., vol. 1, pp. 91–92. Simon is referred to as Rector of Preston in the previous day's notice of his ordination, no doubt in anticipation of the appointment.

[8] Probably about 1300, perhaps in 1292, Simon received the rectorship of Harrow, worth about forty pounds annually, from Archbishop Winchelsey of Canterbury. In any case he had this rectorship in 1303; see *Registrum Ricardi de Swinfield*, ed. Capes, the entry under 23 February 1304, p. 535. On 23 September 1303, Richard of Swinfield, Bishop of Hereford, granted Simon a prebend at Hampton. A prebend is a portion of the revenues of a cathedral granted to a canon as a stipend.

[9] Wolf, *Die Intellektslehre*, p. 26, n. 76. A letter from Edward I, dated 12 April 1306, describes Simon as Doctor of Theology; cf. Great Britain, Public Records Office, *Calendar of Close Rolls, Edward I, (1302–07)*, 12 April 1306, 'To Pope Clement', vol. 2, p. 434.

[10] All of Simon's works were in logic and philosophy, but he did participate in some theological discussions at Oxford around 1300. See Little and Pelster, *Oxford Theology and Theologians*, pp. 335–37, 348–51, for transcriptions of two such theological disputes in which Simon took part.

Simon very likely taught at Paris as well as Oxford.[11] That some of his writings are Parisian is explained by the repeated references to him as 'Simon Anglicus', a name that makes sense only outside England. The colophon to his *Questions on the Prior Analytics* speaks of him as 'magistro Symone Anglico Parisius', and his questions are similar in content and form to Parisian work, especially that of Peter of Auvergne, in the 1270s and early 1280s. He apparently had returned to England by the end of the century since his work seems unaffected by the later Parisian developments of Radulphus Brito's day.[12] Simon assumed the chancellorship of Oxford, the university's highest post, on 31 January 1304.[13] He very likely engaged in little or no scholarly activity after becoming chancellor. This fact and a birthdate of around 1260 would put most of his works in the last two decades of the thirteenth century. Most of the manuscripts containing his works come from the early fourteenth century.[14]

Simon received additional honours the following year, when Archbishop Winchelsey, on 22 September, named him archdeacon of Canterbury, the highest ecclesiastical post in England available to someone not a priest. Rose Graham suggests that Simon received the post because he was not litigious, the archbishop being weary of the continuous disputes in which the previous archdeacon, John de Langton, had been involved. If so, the choice was good, for Simon gave up the position without a fight in November, when the Pope nominated another man for it. As a consolation Simon received the rich rectorship of Reculver, worth 113 pounds annually, but again the Curia had another candidate, and he lost his appointment. But this case he resolved to fight. He had the support of both

---

[11] Simon surely taught at Oxford for most of his career, but he may have been a Master at Paris for a brief period some time before 1300. Ms. Vienna, Wiener Nationalbibliothek, Cod. lat. 2302 describes Simon as 'Magistro Symone Anglico Parysius'; cf. Grabmann, *Die Aristoteleskommentare*, p. 6. Undoubted works of Simon are attributed to 'Simon Anglicus' in several manuscripts. Grabmann concludes that Simon was at Paris as a Master of Arts. There is mention of one 'Symone Anglico' in a Parisian record of 1297, see Wolf, *Die Intellektslehre*, p. 24.

[12] These arguments are from the introduction to Simon of Faversham, *Quaestiones super libro Elenchorum*, ed. Ebbesen et al., pp. 4–5.

[13] Salter, *Snappe's Formulary and Other Records*, pp. 61–64, contains the document confirming his election on this date. The election was always confirmed by the Bishop of Lincoln, or, if we follow the bishop's account of the matter, he granted the office to the candidate proposed by the Oxford Masters. The document transcribed by Salter is Dalderby, Instit., folio 145, and is reported along with an account of the dispute between the Bishop and the Oxford Masters over the Bishop's rights, in Gibson, 'Confirmation of Oxford Chancellors'. At the end of the book, Salter summarizes the conclusions to be drawn from the documents he has recorded. Discussing Simon, he adds to the information he has gleaned from the bishop's confirmations this remark: 'Wood gives John de Osewood as vice-chancellor under Simon of Faversham, on the authority of a deed at University College, but the deed only states that Simon de Faversham committed to John de Osewood the trying of a case' (*Snappe's Formulary and Other Records*, p. 325).

[14] Wolf, *Die Intellektslehre*, p. 20.

archbishop Winchelsey and King Edward I, since the Papal nominee had once been convicted of forgery in England. The archbishop could do little, since he had been suspended from administration by the Pope pending investigation of charges brought against him by the King, but Edward provided Simon with a letter in his favour,[15] and letters of protection for his journey to the Papal Curia.[16] He had already resigned the chancellorship on 9 February 1306 in expectation of his trip.[17] It would be pleasant to report that Simon won his case but, unfortunately, he died shortly after leaving England.[18]

2. Works

Simon's philosophical output can be divided into his logical commentaries and his commentaries on natural philosophy. The commentaries in the first set are almost all edited or being edited; these are:

1) *Questiones libri Porphyrii*; edited by Pasquale Mazarella in Simon of Faversham, *Magistri Simonis Anglici sive de Faverisham Opera Omnia*, vol. 1.

2) *Questiones super librum Praedicamentorum*; edited in Ottaviano, 'Le *Quaestiones super libro Praedicamentorum* di Simone di Faversham', and by Mazarella in Simon of Faversham, *Magistri Simonis Anglici sive de Faverisham Opera*, vol. 1.

3) *Questiones libri Perihermeneias*; edited by Mazarella in Simon of Faversham, *Magistri Simonis Anglici sive de Faverisham Opera*, vol. 1.

---

[15] Great Britain, Public Records Office, *Calendar of Close Rolls, Edward I, (1302–1307)*, 12 April 1306, vol. 2, p. 434, contains the following letter by Edward (the translation is that of the *Calendar*): 'April 12. Wolvesey, to Pope C[lement]. Letter commending to the Pope's favour Master Simon de Faversham, D.D., who is distinguished for learning and virtue, who, obeying the Pope's order, left without making any difficulty the archdeaconry of Canterbury, which he had obtained and of which he enjoyed peaceful possession for some time, by reason of the collation thereof made by the Pope upon the son of Amaneus de Lebreto, and afterwards having obtained the church of Raycolvre, in the diocese of Canterbury, he fears that he will be vexed and disquieted by reason of a provision for a clerk, who is much his inferior in merits and goodness, that is said to have emanated from the Pope's authority in that Diocese, and requesting the Pope, after weighing the premisses, to favour Simon so that he may receive in effect what the king has suggested'. For the value of Reculver, see Little and Pelster, *Oxford Theology and Theologians*, p. 263.

[16] Great Britain, Public Records Office, *Calendar of Close Rolls, Edward I, (1302–1307)*, 24 May 1306, vol. 2, p. 435: 'simple protection, until Easter, for Master Simon de Favresham, going to Rome'.

[17] Salter, *Snappe's Formulary and Other Records*, pp. 64–66. Transcription of Dalderby, Instit., folio 149, the Bishop of Lincoln's confirmation of Simon's successor to this office.

[18] In *Registrum Roberti Winchelsey*, ed. Graham, vol. 2, pp. 1044–46, a letter of 1309 (undated, but it refers to events of 1308) describes Simon, one-time rector of Reculver, as deceased.

4) *Questiones super librum Priorum*. Manuscripts: Milano (=M), B. Ambros., C. 161 inf., ff. 34v–64v; Oxford (=O), Merton College 292, ff. 111–137v; Paris, BnF, lat. 16125, ff. 37–60v; Prague, Metr. Kap., L. 66, ff. 1–30v; Vienna, NB, 2302, ff. 16–23v.

5) *Questiones super librum Posteriorum, redactio 1*. Manuscripts: Kassell (= K), Murhardsche Bibliothek der Stadt, Cod. 2o Ms. Phys. 11, ff. 78r–99r; M, ff. 79v–99r; O, ff. 138ra–156vb.[19]

6) *Questiones super librum Posteriorum, redactio 2*. Manuscript: M, ff. 99r–112r. Edited in the present work.

7) *Dicta super librum Topicorum*. Manuscript Leipzig, UB 1359, ff. 24r–43ra, attributed to 'magister Simonis'.

8) *Questiones super librum Elenchorum, redactiones 1 et 2*; edited by Sten Ebbesen et al. in Simon of Faversham, *Quaestiones super libro Elenchorum*.

9) *Questiones super universalia*. Manuscript: Kassel, LB, 2° Philos., ff. 23v–26r.

10) *Sophisma 'universale est intentio'*; edited in Yokoyama, 'Simon of Faversham's *Sophisma*', and revised in Pinborg, 'Simon of Faversham's *Sophisma*'.

11) *Commentarius in Petri Hispani Tractatus*. Manuscripts: Milano, B. Ambros. F 56 sup.; München, CLM 14697, ff. 1–48; Padova, Anton. 429 Scaff. XX, ff. 1–48; Paris, BNF lat. 16126, ff. 77–90 (incomplete). Attributed to Simon by Sten Ebbesen.[20]

The following commentaries on natural philosophy are almost certainly by Simon:

1) *Questiones super librum Physicorum*. Manuscripts: O, ff. 185r–239v; Erfurt, Bibliotheca Amploniana F348, ff. 1r–69r; San Juan, Casa del libro, ff. 2r–24v. For this one work an approximate date can perhaps be ascertained: it is a *reportatio* by Robert of Clothale, and it can be argued that it could not have been done before 1290. The questions are listed in Zimmermann, *Verzeichnis Ungedruckter Kommentare*, pp. 190–96.

---

[19] Edited by John Longeway in Longeway, *Simon of Faversham's Questions on the Posterior Analytics*.
[20] See Weijers and Calma, *Le travail intellectuel*, p. 107.

2) *Questiones super librum De anima*. The manuscript situation is complex; see Weijers and Calma, *Le travail intellectuel*, p. 109. The partial version in O, ff. 364rb–370vb has been edited in Sharp, 'Simonis de Faverisham *Quaestiones*' (only questions on Book III). Vennebusch, 'Die *Quaestiones in tres libros*', pp. 20–39, lists all the questions.

3) *Recollectiones super librum Meteororum*. Manuscript: Oxford, Bodleian Library, Tanner 116, ff. 68–82.

4) *Questiones super librum De motu animalium*. Manuscript: O, ff. 393va–396vb; partially edited in Christensen, 'Simon of Faversham *Quaestiones*'.

5) *Questiones super Parva naturalia*:

   a) *Questiones super librum De somno et vigilia*. Manuscript: O, ff. 389ra–393vb; edited in Ebbesen, 'Simon of Faversham. *Quaestiones*'.

   b) *Questiones super librum De longitudine et brevitate vitae*. Manuscript: O, ff. 396vb–399ra; partially edited in Ebbesen, 'Gerontobiologiens Grundproblemer'.

   c) *Questiones super librum De iuventute et senectute*. Manuscript: O, ff. 399ra–400r.

   d) *Questiones super librum De inspiratione et respiratione*. Manuscript: O, ff. 400r–401va.

3. Manuscript M[21]

M: Milano, Biblioteca Ambrosiana, cod. C.161.inf.

A description of this manuscript is found in Shooner, *Codices Manuscripti operum*, pp. 330–32, and in Simon of Faversham, *Quaestiones super libro Elenchorum*, ed. Ebbesen et al., pp. 15–16. It is a composite manuscript, but the first 116 leaves belong together. These contain Simon of Faversham's question commentaries on Porphyry's *Isagoge*, and on Aristotle's *Categories, De interpretatione, Prior Analytics, Posterior Analytics*, and *Sophistici Elenchi*, with Siger of Brabant's *Impossibilia* and some minor works of Thomas Aquinas. They are written in black ink, with paragraph markings (large Cs with slashes through them, in red), by a continental scribe from around the end of the thirteenth century. The hand is professional and quite legible.

[21] This account is also but a restructuring of John Longeway's description of ms. **M** in the introduction to his editions of his commentaries on *Posterior Analytics*. The rightful author of this part is also John Longeway.

The *Quaestiones super librum Posteriorum, redactio 1*, occupy folios 79v to 99r, beginning a little below the middle of the first column with an elaborate capital S, seven lines in height, preceded by the title 'Quaestiones libri Posteriorum' and, in another hand and brown ink, 'Secundum M. Simonem✝'. They follow immediately on the *Quaestiones super librum Priorum*. The incipit is: *Sicut dicit philosophus in X Ethicorum*. There is no explicit, but there is also no reason to think that any of the work has been omitted.

The *Quaestiones super librum Posteriorum, redactio 2*, here edited, occupy folios 99r to 112v,[22] beginning about twenty lines down in the second column, immediately after *redactio 1*, in the same hand, with an elaborate capital O, 5 lines in height. There is no title, but the same secondary hand in brown ink adds in the margin: *Incipit quaestiones secundi Posteriorum secundum dictum M. Simonem Anglicum ✝*. The incipit is: *Omnis doctrina et omnis disciplina etc.*, this much written in larger print, as are most of the lemmata from *Posterior Analytics* both here and in the first set of questions. The last line of these questions on folio 112va has extra spaces added to stretch out the words 'patet ad rationes' to the right margin. The rest of the column is then blank with the explicit, again in the secondary hand: *Expliciunt Quaestiones libri Posteriorum secundum M. Simonem Anglicum bene dicte per universa studia ✝*.

4. The text transmitted by M

The version edited below is made up of 69 questions, 42 on Book I, 27 on Book II. The questions are quite short, the arguments concise, and the style bears many marks of orality.

The quality of the text transmitted by M is rather poor. It contains all kinds of errors: bad lectures, *lapsus*, omissions, repetitions etc. Presumably, the copyist had a very imperfect model in front of him. For example, question 41, the penultimate in Book I, ends *ex abrupto* in the middle of the *solutio*: 'et ideo corrupta re quantum ad esse extra animam non corrumpitur necessario scientia de ipsa. Verumtamen ...', so the end of the *solutio* and the response to the arguments are missing. We find five *lacunae*, viz. blanks, indicating that the copyist was unable to read his model, or that the model was left unfinished (cf. p. 178, l. 38; p. 180, l. 13; p. 182, l. 36). In some cases, the text was so corrupted that we were unable to propose an acceptable correction (cf. p. 220, l. 50–51). Most of the errors in M are fairly trivial, however, and were easily rectified.

[22] The digital reproduction of the manuscript M is available at https://digitallibrary.unicatt.it/veneranda/0b02da828007c5bd. The marginal pagination markers in the present edition contain links to specific folios.

It is noteworthy that some sentences appear to be glosses that have been inserted into the main text of M, as in the following case (p. 30, l. 17–22):

> Unde addiscens aliquam conclusionem prius scivit eam in universali et in suis principiis (secundum enim Lincolniensem principia habent virtutem activam et universalem super conclusionem), tamen non prescivit eam in particulari. Unde addiscens conclusionem scit eam prius in universali et in virtute, tamen ignorat in particulari.

Here, the words in brackets clearly break the flow of the text and look like a note added by a reader. Other similar examples can be found in the rest of the text (p. 90, l. 151–154).

5. Sources

As usual in Parisian commentaries of his time, Faversham's sources are mostly ancient, late ancient, and medieval authorities. These authorities are Greek, Roman, Arabic, Hebraic, and medieval Latin.

Greek and Roman sources

Obviously, the most solicited Greek authority is Aristotle, with many references to the whole *Organon*, and to *Physics*, *De caelo*, *De generatione et corruptione*, *De anima*, *Historia animalium*, *Generatio animalium*, and *Metaphysics*. Additionally, there is one reference to Ammonius' commentary on *De interpretatione* (q. 28), one to Porphyry's *Isagoge* (q. 36), one to Proclus' *Elementatio theologica* (q. 9), and five to Themistius, one to his *De anima* (q. 36), and four to his paraphrase on *Posterior Analytics* (q. 4, 28, 30, 35). Finally, there are also two references to the pseudo-Aristotelian *Liber de causis* (q. 6 and 36). There are references to Theophrastus, but in all likelihood these are based on Simon's acquaintance with Albert the Great's paraphrases on Aristotle.

The only Roman source is Boethius, with references to *De hebdomadibus* (q. 49), *De topicis differentiis* (q. 36), and *De trinitate* (q. 49).

Arabic and Hebraic sources

The Arabic sources are mostly Avicenna and Averroes. For Avicenna, the references are to his *Logica* (q. 21 and q. 58), *De philosophia prima*

(q. 4, 18, 21, 22, 41, 45, 49, 58, 59) and *Sufficientia* (q. 13). For Averroes, the references are to his *De anima* (q. 8, 22, 41), *Metaphysica* (q. 6, 11, 29, 38, 44, 49, 59), and *Physica* (q. 6, 48, 65, 66). Noteworthy are also his three references to al-Ghazali's *Logica* (q. 22, 49, 58). Finally, there is one reference to Maimonides' *Guide for the Perplexed* in q. 36.

Medieval Latin sources

The medieval Latin authorities to which Simon refers are Robert Grosseteste, Albert the Great, Thomas Aquinas and Giles of Rome. Grosseteste, the *Lincolniensis*, is used as an authority on *Posterior Analytics*: the references are to I.1 in q. 5 and 8, to I.4 in q. 12 and 16, twice to I.7 in q. 40, and to I.8 in q. 26. The references to Albert the Great are to his paraphrase on *Liber de causis*, to I.I.8 and II.I.9 in q. 49, and to that on *Posterior Analytics*, to I.I.3 in q. 5, I.I.5 in q. 1, I.II.5 in q. 9, II.I.1 in q. 43 and 45, II.I.2 in q. 45, II.I.3 in q. 46, II.II.10 in q. 31 and II.III.7 in q. 15.

Faversham's use of Thomas Aquinas and Giles of Rome goes beyond their exposition and commentary on *Posterior Analytics*, for their positions in metaphysics and natural philosophy are also brought to bear on Simon's discussion. Giles of Rome's commentary on *Posterior Analytics* II.9 is referred to in q. 31, where Simon also refers to Giles' *De medio demonstrationis*. In q. 41 there is one reference to Giles' commentary on *De generatione* I.1 and in q. 49 one reference to *Theoremata de esse et essentia* (t. XXII) and another to *Questiones disputatae*, q. 9. Thomas Aquinas' exposition on *Posterior Analytics* is referred to in q. 34 (to II.1) and in q. 68 (to II.9); in q. 49 there are two references to Aquinas, one to *De ente et essentia* IV and another to *De potentia dei* q. 5 ar. 4.

6. Content

Simon of Faversham's *redactio secunda* on *Posterior Analytics* consists of sixty-nine questions, of which questions 1 to 42 concern Book I and 43 to 69 concern Book II. The commentary does not have a *prooemium* and tackles Aristotle's *littera* from the outset, with a first question asking whether Aristotle's opening claim in *Posterior Analytics*: *omnis doctrina et omnis disciplina intellective ex preexistente fit cognitione* (I, 1, 71 a 1, tr. Iacobus Veneticus), with question two asking whether there can be a science of demonstration.

At any rate, questions 1 and 3 to 8 tackle *Posterior Analytics* I.1 specifically in relation to the different epistemic issues raised by Meno's paradox.

Questions 9 to 14 turn to I.2, where the issues discussed are related to Aristotle's definition of the demonstrative syllogism.

Question 15 deals with the claim in I.3 to the effect that there is not demonstration about every kind of thing.

Questions 16 to 37 turn to the characterisation of per se and universal attribution in *Posterior Analytics* I.4 and their different kinds. It is noteworthy that some of these questions deal with assertions that are usually discussed in the sophismatic literature of the period, such as the assertions 'animal est homo', 'animal est rationale', 'homo est animal homine non existente' and 'Sortes est homo'.

Questions 38 and 39 deal with Aristotle's discussion, in I.5, on the proper universal attribution required in demonstration.

Question 40 discusses Aristotle's claim, in I.6, that contingent premises might produce a necessary conclusion.

Questions 41 and 42 are concerned with the possibility of having demonstration of corruptible things, given Aristotle's claim that the attribution targeted by the conclusion must be eternal.

Question 43 opens with the lemma: *Questiones sunt equales numero etc.* (II, 1, 89 b 24) pointing to the first line of II.1 and asks whether there are as many questions as things that can be known scientifically.

Questions 44 to 46 continue this line of inquiry specifically in relation to the questions *si est, quid est* and to Aristotle's claim that all questions ask for the middle term.

Questions 47 to 61 turn to the thorny issue of the limits between definition and demonstration and the related issue about the demonstration and the definition of the essence, which Aristotle discusses from II.3 to II.7. Question 49 brings the medieval distinction between essence and existence to bear on Aristotle's suggestion, in II.7, that in demonstration the existence of the subject is presupposed.

Question 62 opens with the lemma: *Iterum autem speculandum* (II, 8, 93 a 1), which points to the first line of II.8. This question asks whether a thing can be its own cause. The role of the cause in demonstration continues to be discussed in questions 63 to 69 in relation to Aristotle's claims about causes in II.8 to II.11. Simon's commentary stops short of commenting on *Posterior Analytics* II.12 to 19.

7. Principles of edition

This work took John Longeway's edition (1977) as its point of departure. We have checked his edition against the manuscript and revised it as necessary. We have tried to respect the oral, unpolished, nature of the text, so we have intervened only in the following cases. First, grammatical incongruencies which are not already common in a text of this kind; so, for instance, the indicative has been kept in places where the subjunctive would have been the correct option, or the present where the future would have been expected in a more polished, edited, text. Second, readings that are clearly wrong given the content that should be expected given that we are working with a commentary on *Posterior Analytics*. Third, we have corrected textual environments in some few places where without the proposed corrections the piece of text would have been very difficult to understand.[23] In all three types of correction, we have at times bracketed out (within [...]) letters, words, or pieces of text that are clearly mistakenly added, and inserted (within ⟨...⟩) letters, words, or pieces of text that are mistakenly omitted, in some cases by homeoteleuton.

We have added numbers—1.1, 2, 3, Ad 1.1 etc.—in order to make clear to the reader the dialectical structure of the argument. The numbers indicate which are the opening arguments (1), the arguments in oppositum (2), Simon's solution (3) and his answers to the opening arguments (Ad 1 etc.). In several places, there are also answers to the arguments in oppositum, marked as Ad 2 etc. This numbering is added for ease of reading and, of course, is not part of Simon's Latin text.

The critical apparatus also indicates the corrections made by the medieval scribe. Ancient, late ancient, and medieval authoritative sources referred to by name are, whenever possible, identified and included in the apparatus. Finally, whenever possible, we have separated the premises of syllogistic arguments with semicolons.

We have added our own punctuation and have kept the medieval use of 'e' instead of the classical diphthong 'ae'. However, we have uniformized words with idiosyncratic spelling in the manuscript.

Generally, our treatment of the text is in line with John Longeway's principles for the edition, which was the starting point of this final version.

---

[23] This mainly thanks to the invaluable suggestions by Cecilia Trifogli after her careful assessment of a first draft of this edition. We are immensely thankful to Cecilia for her conscientious help.

8. The translation

The English translation that accompanies this edition is based on the one that appears in John Longeway's dissertation. It has been significantly revised by Matthew Wennemann in light of the revised text of the Latin that appears in this volume.

Abbreviations

Critical Apparatus

a.c.	ante correctionem
add.	addidit
corr.	correxit
def.	deficit
dub.	dubius
exp.	expunxit
hom.	homeoteleuton
inv.	invertit
i.m.	in margine
iter.	iteravit
lac.	lacuna
om.	omisit
p.c.	post correctionem
s.l.	supra lineam
ut v.	ut videtur
⟨...⟩	addidi
[...]	delevi
***	spatium vacans

Apparatus of Sources: works of Aristotle

AnPo	Analytica posteriora
AnPr	Analytica priora
Cat.	Categoriae
De an.	De anima
Sens.	De sensu
Eth. Nic.	Ethica Nicomachea
Met.	Metaphysica
Phys.	Physica
SE	Sophistici elenchi
Top.	Topica

Questiones super librum Posteriorum
redactio secunda

| Incipiunt questiones libri *Posteriorum* secundum dictum magistrum Simonem Anglicum etc.

M 99rb

⟨Liber I⟩

Omnis doctrina et omnis disciplina etc. (71 a 1–2)

⟨Questio 1⟩

Circa librum *Posteriorum* queritur utrum ista propositio sit vera: *omnis doctrina* etc.

1.1 Et videtur quod non, quia sicut est de cognitione sensitiva, sic de cognitione intellectiva; sed doctrina sensitiva non fit ex preexistente cognitione; ergo etc.

1.2 Item, multa sunt que sunt naturaliter nota; ergo non fiunt ex preexistente cognitione. Antecedens patet, quia principia sunt talia.

2 Oppositum patet per Philosophum hic.

3 Ad hoc est intelligendum, secundum quod dicit Alexander[a] hic, quod duo sunt modi cognitionis: cognitio per inventionem et per disciplinam; unde intentio Philosophi secundum illum est *omnis doctrina* etc., id est, ex preexistente inventione. Sed hoc imponitur Alexandro, et si hoc dixit, male dixit quod omnis cognitio nostra fit ex inventione priore. Si tunc

---

[a] Cf. Albertus Magnus, *In Analytica posteriora*, I, I, 3 (ed. Borgnet, p. 8); Albertus habet 'antiquis Peripateticis'

---

incipiunt questiones ... etc.] *add. i.m. ab alia manu* M   12 hoc¹] hic *ante corr.* M ||
alexandro] alexander M   13 quod] quia M

Ed. ©Costa et al. CC BY-NC 4.0
Trans. ©Longeway and Wennemann, CC BY-NC 4.0

Questions on the Posterior Analytics
Second Redaction

Here begins the questions on the *Posterior Analytics* according to Master Simon the Englishman.

Book I
'All teaching and all intellectual learning...'

Question 1.

Concerning the *Posterior Analytics* it is asked whether this proposition is true, 'all teaching and all intellectual learning come from preexisting cognition'.

1.1 And it seems that it is not, for as it is with sensory cognition, so it is with intellectual cognition. But sensory teaching does not arise from preexisting cognition. Therefore etc.

1.2 Furthermore, there are many things which are known naturally; therefore they do not arise from preexisting cognition. The antecedent is obvious, since first principles are of this sort.

2 The opposing view is clearly true from what the Philosopher says here.

3 In response to this, it should be understood, in accord with what Alexander says here, that there are two manners of cognition, cognition through discovery and through learning. Hence the meaning of the Philosopher, according to Alexander, is, 'all teaching' etc., is from preexisting discovery. But Alexander imposes this interpretation on the text, and if the Philoso-

omnis cognitio nostra fit ex inventione priore, prima cognitione esset alia cognitio prior, vel si aliquid addisceremus, infinita prius addisceremus.

Unde dicendum quod omnis doctrina fit ex precognitione intellectiva, et propter hoc hic declaranda sunt duo: primum est quod omnis doctrina fit ex preexistente cognitione, secundum, quod ex cognitione intellectiva. Probatio primi, quia nihil aliud intelligitur per hoc quod omnis doctrina et omnis disciplina fit ex preexistente cognitione nisi quod omnis conclusio fit ex cognitione preexistente; conclusio autem est nomen alicuius ignoti, et in cognitionem ignoti devenimus per aliquod notum.

Item, ratio Theophrasti[a] est ad hoc: omne quod movetur habet aliquid illius ad quod movetur; si ergo debemus cognoscere conclusionem, oportet prius habere aliqualem cognitionem de conclusione.

Item, probatio secundi, quia cognitio conclusionis est ex cognitione principiorum, cognitio principiorum est [ex] cognitio[ne] intellectiva. Et ulterius cognitio principiorum est ex cognitione terminorum, cognitio terminorum est ex cognitione sensitiva. Unde competenter dixit Philosophus[b] quod omnis doctrina et omnis disciplina fit ex preexistente cognitione sensitiva, et ibi est status.

Ad rationes.

Ad 1.1 Ad primam, dico quod in hoc est similitudo quod sicut sensus est ⟨prius⟩ in potentia et posterius in actu, sic et intellectus. Sed dissimilitudo est in hoc, quia cognitio intellectiva est ex priore, sensitiva autem non. Item, dissimilitudo est in hoc, quia cognitio intellectiva causatur a sensitiva, sed non e converso.

Sed aliquis argueret in contrarium: posterius non est causa prioris; sed sensus est posterior intellectu; ergo etc. Dico quod 'prius' dicitur dupliciter: aut secundum substantiam et perfectionem aut secundum generationem. Unde sensus est prior secundum generationem, intellectus secundum substantiam et perfectionem.

[a] Cf. Albertus Magnus, *In Analytica posteriora*, I, I, 5 (ed. Borgnet, p. 18)   [b] Cf. Aristoteles, *AnPo*, II, 19, 99 b 15–100 b 17

---

17 doctrina] cognitio M   20 disciplina] *i. m.* M   ‖ conclusio] cognitio *ante corr.* M   34 posterius] primo M   38 argueret] *iter.* M

pher said this, he spoke badly, in saying that our every cognition arises from prior discovery. For if every cognition arises from prior discovery then there must be another cognition prior to the first cognition, or else if we have learned anything at all then we have already learned an infinite number of things.

Hence it should be said that all teaching arises from intellectual precognition, and because of this two things should be explained. The first is that every cognition arises from preexisting cognition, and the second is that every teaching arises from intellectual cognition. The proof of the first is that nothing else is understood by this statement that all teaching and all learning arise from preexisting cognition than this, that every conclusion arises from preexisting cognition. But a conclusion is a name for something unknown, and we arrive at a cognition of the unknown through something known.

Furthermore, the argument of Theophrastus is to this effect. Everything which is moved has something of that toward which it is moved; if then we are to cognize a conclusion we must already have some cognition about the conclusion.

Furthermore, the proof of the second: The cognition of a conclusion is from the cognition of principles, the cognition of principles is from intellectual cognition. Further, the cognition of principles is from the cognition of terms, and the cognition of terms is from sensory cognition. Hence, the Philosopher rightly said that all teaching and all learning arise from preexisting sensory cognition, and that is the end of the series.

In response to the arguments.

Ad 1.1 In response to the first argument I maintain that there is a similarity here because just as sense is prior in potentiality, and posterior in actuality, so also is the intellect. But there is a dissimilarity in that intellectual cognition is from something prior, but sensory cognition is not. There is also a dissimilarity in that intellectual cognition is caused by sensory cognition, but not conversely.

But someone might argue to the contrary: the posterior is not the cause of the prior, but sense is posterior to intellect, therefore etc. I maintain that 'prior' is said in two ways, either according to substance and perfection, or according to generation. Hence, sense is prior according to generation, and intellect according to substance and perfection.

Ad 1.2 Ad aliam, dico quod aliquid esse naturaliter notum est dupliciter: aut quia aliquid sit in substantia anime nostre, et sic nihil est nobis naturaliter notum, quia tunc non esset anima nostra sicut tabula rasa; aut quia statim intellectus noster est in potentia ad intelligendum illa, et talia sunt principia prima. Unde ⟨cognitio⟩ illorum non est ex cognitione priore intellectiva, sed ex cognitione priore sensitiva.

⟨Questio 2⟩

Consequenter queritur utrum de demonstratione possit esse scientia.

1.1 Et videtur quod non, | quia si de demonstratione sit scientia, cum scientia sit per demonstrationem, quero tunc de illa demonstratione utrum de ipsa sit scientia aut non; si non, ergo nec de prima, si sic, queram de illa, et sic in infinitum.

1.2 Item, aliquis non habet scientiam de conclusione nisi prius sciat ⟨se habere⟩ demonstrationem ad illam conclusionem. Si ergo prius sciat se habere demonstrationem ad illam conclusionem, prius scivit se habere demonstrationem ad illam demonstrationem, ergo prius demonstrationem habet quam sciat demonstrationem; hoc autem est impossibile; ergo etc.

2 Oppositum patet per Philosophum hic.

3 Ad questionem dico quod de demonstratione potest esse scientia, et iterum de demonstratione est scientia, et iterum quod de demonstratione est scientia per demonstrationem.

Probatio primi, quia de omni eo quod est intelligibile ab intellectu et habet proprietates et passiones potest esse scientia; sed demonstratio est huiusmodi.

---

2 de] in M 6 sciat] habeat *ante corr.* M 16 passiones] non *add. sed exp.* M

Ad 1.2 In response to the other argument, I maintain that there are two ways in which something is known naturally, either because something is in the substance of our soul, and in this way nothing is naturally known by us since then our soul would not be like a blank tablet. Or something can be naturally known because our intellect has an immediate potential to understand it, and first principles are of this sort. Hence, the cognition of those things is not from a prior intellectual cognition, but a prior sensory cognition.

Question 2.

Next it is asked whether it is possible for there to be a science of demonstration.

1.1 It seems it is not, for if there were a science of demonstration, since science occurs through demonstration, I ask concerning this demonstration whether there is a science of it or not. If not, then neither is there science of the first; if there is, I will ask about that science as before, and so on ad infinitum.

1.2 Furthermore, one does not have scientific knowledge of a conclusion unless he first knows himself to have a demonstration for that conclusion.[1] So if he first knows himself to have a demonstration for that conclusion, then before that he knew himself to have a demonstration for that demonstration. Therefore he has the demonstration before he has scientific knowledge of it. But this is impossible. Therefore etc.

2 The opposite view is obvious through what the Philosopher says here.

3 I reply to the question that, first, there can be a science of demonstration; second, there is a science of demonstration; third, there is a science of demonstration through demonstration.

Proof of the first point: There can be a science of anything the intellect can understand and that has properties and attributes. But demonstration is like that.

---

[1] Here and throughout, 'know(s)' almost always translates forms of 'scire' (occasionally, 'notus' and its forms are translated by the word 'known'). It therefore has the technical meaning of knowing scientifically, i.e., knowing something by demonstration from the first principles of a body of knowledge. 'Scientia' is rendered as 'knowledge' or, where it is especially useful to bring out this technical meaning, 'science'.

Probatio secundi, quia de eo quod est commune cuilibet scientie speciali, de eo est aliqua scientia; sed demonstratio est commune cuilibet scientie speciali; ergo etc. Et illa scientia est communis, non specialis, quia qua ratione determinatur in una, et in alia, et ideo determinatur in scientia communi, cuiusmodi est logica. Sed cum logica sit in aliqua parte de demonstratione, in aliqua parte de syllogismo probabili, intelligendum est quod quoad illam partem que est de demonstratione est tantum docens et non utens, quia procedit ex communibus, et demonstratio in scientiis specialibus non est ex communibus. Sed quantum ad illam partem aliam est docens et utens: docens quia docet ex quibus et qualibus sit arguendum probabiliter, utens quia operatur in aliis scientiis.

Probatio tertii, quia omnis scientia tradit doctrinam per demonstrationem. Cum ergo logica sit scientia, tradit de demonstratione scientiam per demonstrationem. Et hoc scitur per demonstrationem eandem specie et non numero. Et sicut non est inconveniens procedere in infinitum circulariter in generatione, sic nec in demonstratione, sicut est in operatione intellectus intelligentis alia, et intelligendo alia intelligit se [intelligere se], et postea intelligit se intelligere se, sic in infinitum.

Ad rationes.

Ad 1.1 Ad primam, concedo quod demonstratio fit per demonstrationem, nec est inconveniens ibi processus in infinitum. Vel si non contingat procedere in infinitum in demonstratione, potest demonstratio prima accipi ab aliquo sensu, sicut possumus de ista regula 'omne totum est maius sua parte', 'et domus est quoddam totum, ergo etc'. – omnia ista possumus sensu comprehendere. Unde si demonstratio fuisset adgenerata in mente, tamen credo quod illam habuit per inventionem.

Ad 1.2 Ad aliam, dico quod duplex est scientia: una in qua differt scientia et modus sciendi, sicut est in omnibus scientiis specialibus, alia in qua non differt scientia et modus sciendi, cuiusmodi est logica, cum ipsa sit suus modus sciendi; simul enim scit demonstrationem et modum demonstrandi. Unde logicus, licet sciat demonstrationem, non oportet quod sciat se habere demonstrationem ad illam, sed maior solum intelligenda est in scientiis specialibus.

Proof of the second: There is a science of that which is common to every special science; but demonstration is common to every special science; therefore etc. And this science is common, not special, since it is decided by the same argument in one science and another, and so it is decided in a common science to which kind logic belongs. But since logic is in one part about demonstration and in another part about probable syllogisms, it should be understood that as far as that part which is about demonstration is concerned, it is merely a teaching science, and not a science in use, since it proceeds from common things, and demonstration in the special sciences is not from common things. But insofar as the other part is concerned, it is both a teaching science and a science in use. It is a teaching science since it teaches from which and what sort it is to be argued probably, and it is science in use because its operation occurs within the other sciences.

Proof of the third point: Every science hands down some teaching through demonstration. Since logic is a science it passes on its scientific knowledge about demonstration through demonstration, and this is known through a demonstration which is the same in species but not the same in number. And as it is not absurd to proceed to infinity in a circular manner in generation, so also it is not absurd in demonstration. So it is in the operation of the intellect when it understands other things: in understanding those things it understands itself, and afterwards it understands that it understands itself, and so on to infinity.

In response to the arguments.

Ad 1.1 In response to the first I grant that demonstration arises through demonstration, nor is it absurd that there should be a procession to infinity there. Alternatively, if one cannot proceed to infinity in demonstration a first demonstration can be received in some sense, as we can receive a demonstration about this rule: every whole is greater than its part, and a house is a certain whole, therefore etc. All of this can be grasped by the senses. Hence, even if a demonstration has been produced in the mind, I nevertheless believe that one had it through discovery.

Ad 1.2 I reply to the second objection that there are two sorts of sciences. In one sort the science and its way of knowing differ, as for instance in all special sciences. In the other sort the science and its way of knowing do not differ, and logic is like this, since it is its own way of knowing. For one knows demonstration and how to demonstrate simultaneously. Hence, a logician may know a demonstration, and there is no need for him to know scientifically that he has a demonstration for it. But the major premise is only to be understood as applicable to the special sciences.

⟨Questio 3⟩

Consequenter queritur circa partem illam: *dupliciter autem oportet precognoscere* etc. (71 a 11–17), utrum tantum due sint precognitiones secundum quod dicit, scilicet quid est et quia est. Secundo utrum definitio indicans quid est res et quid significat nomen sint una et eadem.

1.1 De primo arguitur quod non, quia omnis questio est quedam precognitio, quia omne quod queritur aliquo modo precognoscitur, aliter numquam inveniretur; sed quattuor questiones ⟨sunt⟩, sicut vult Philosophus[a] secundo huius; ergo etc.

1.2 Item, cuilibet precognito respondet sua precognitio; sed tria sunt precognita, scilicet subiectum, passio et dignitas; ergo etc.

1.3 Item, sicut contingit precognoscere quid est, ita contingit precognoscere quale est et quantum est et sic de aliis; ergo etc.

2 Oppositum vult Philosophus.

3 Ad questionem dico quod tantum sunt due precognitiones. Et secundum quod vult Philosophus[b] secundo huius quattuor sunt questiones, scilicet 'si est', 'quia est', 'quid est' et 'propter quid est', et ideo illa que vere scimus quattuor sunt, scilicet dicta. Ex hoc arguitur quod tantum sunt due precognitiones, quia de numero illorum scibilium illa solum precognoscuntur ante demonstrationem quorum cognitio per demonstrationem non habetur. Sed quid est subiectum, et quid significat nomen subiecti, et quia est subiectum, et quid est quod significat passio, et quia est dignitas per demonstrationem non habetur. Ergo tantum sunt due precognitiones, quia sola ista cognoscuntur ante demonstrationem. Probatio minoris, quia cognitio quidditativa subiecti est causa omnium passionum que possunt demonstrari de subiecto, et si illa est causa, per demonstrationem non habetur, quia est ante omnem demonstrationem. De passione similiter oportet scire quid est, quia nihil potest demonstrari de aliquo

---

[a] Aristoteles, *AnPo*, II, 1, 89 b 24–25   [b] Aristoteles, *AnPo*, II, 1, 89 b 24–25

7 quattuor] communi M

Question 3.

Next we ask about the phrase, 'Now it is necessary to cognize in advance in two ways, etc.', whether there are two advance cognitions as he says, namely what-it-is and that-it-is-so, and in the second place, whether the definition indicating what a thing is and what the name signifies are one and the same.

1.1 Concerning the first it is argued that there are not two advance cognitions, for every question is a kind of advance cognition, for everything that is asked can in some way be cognized in advance—otherwise it would never be discovered. But there are four questions, as the Philosopher has it in the second book of this work. Therefore etc.

1.2 Furthermore, to whatever is cognized in advance there corresponds its advance cognition. But three things are cognized in advance: the subject, the attribute, and the axiom. Therefore etc.

1.3 Furthermore, just as there is an advance cognition of what-it-is, so also there is an advance cognition of what sort it is, and how much, and so on for the others. Therefore etc.

2 The Philosopher takes the opposite view.

3 I reply to the question that there are only two advance cognitions, and as the Philosopher has it in the second book of the *Posterior Analytics*, there are four questions: if-it-is, whether-it-is-so, what-it-is, and because-of-what-it-is.[2] And therefore there are four things that we truly know, namely the aforesaid. From this, it is argued that there are only two advance cognitions, since among what is scientifically knowable those alone are cognized in advance, prior to demonstration, whose cognition is not grasped through demonstration. But what the subject is and what its name signifies, and that the subject is, and what it is that the attribute signifies, and that something is an axiom, these are not grasped through demonstration. Therefore there are only two advance cognitions, since these alone are cognized before demonstration. Proof of the minor: since quidditative cognition of the subject is the cause of every attribute which can be demonstrated of the subject, and if that is the cause, it is not

---

[2] These four questions translate the Latin phrases *si est*, *quia est* (sometimes rendered as 'that-it-is-so'), *quid est*, and *propter quid est* (often rendered as 'because-of-what' or 'because-of-which'). Where they appear in this usage, as one of these four questions or the corresponding cognitions of the answers, they are hyphenated. Occasionally, where the questions have a grammatical subject and the technical usage is clear (as immediately below), they are unhyphenated.

nisi sciatur quid est; cum ergo de passione debeat aliquid probari, oportet precognoscere quid est. Et similiter de dignitate.

Sed et tu dices: videtur quod debeamus habere tres precognitiones, | scilicet quid est subiectum secundum suam substantiam, alia: quid significat nomen, tertia: quia est. Dicendum quod una istarum precognitionum reducitur ad aliam, scilicet illa que est secundum substantiam et naturam subiecti ad illam que est quid significatur per nomen; numquam enim sciretur quid significaretur per nomen nisi sciretur que est substantia rei. M 99vb

Ad rationes.

Ad 1.1 Ad primam, dico quod non ex eodem sumitur numerus questionum et precognitionum, quia numerus questionum sumitur ex his que vere [scimus] sciuntur per demonstrationem, cuiusmodi sunt quattuor; numerus precognitionum sumitur ex his que sciuntur preter demonstrationem, et talia sunt duo tantum: quid est et quia est.

Ad 1.2 Ad aliam, dico et concedo quod tres sunt precognitiones secundum numerum, scilicet subiectum, dignitas et passio, tamen reducuntur ad duas: ad complexum et incomplexum, et talia precognoscuntur: quid est et quia est.

Ad 1.3 Ad tertiam, dico quod [de aliquo] non potest precognosci aliquid esse quale vel quantum, quia intelligere aliquid esse quale est intelligere qualitatem sibi inesse, et qualitatem sibi inesse scimus per demonstrationem, et nullum scitum per demonstrationem precognoscitur; et ideo non possumus precognoscere aliquid esse quale vel quantum.

---

40 sciuntur] sumuntur M   43 subiectum] pa *add. sed del.* M

grasped through demonstration, since it is before every demonstration. In the same way, it is necessary to know concerning the attribute what it is, since nothing can be demonstrated of anything unless we know what it is. Since, therefore, something must be proved of the attribute, it is necessary to cognize in advance what-it-is, and similarly of an axiom.

But you will say, it seems that we ought to have three advance cognitions, namely what the subject is according to its substance, second, what the name signifies, and third, that it is. It should be said that one of these advance cognitions is reduced to the others, namely that which is according to the substance and nature of the subject to that concerning what it is that is signified by the name. For it is never known what is signified through the name unless it is known what the substance of a thing is.[3]

To the arguments.

Ad 1.1 To the first I reply that the number of questions and the number of advance cognitions are not taken from the same thing, for the number of questions is taken from those which we truly know through demonstration, of which sort there are four; but the number of advance cognitions is taken from those things which are known outside demonstration, and such are only two, what-it-is and that-it-is-so.

Ad 1.2 To the other, I reply and grant that three things in number are cognized in advance, namely subject, axiom, and attribute, but they are reduced to two, to the proposition and the term, and such questions are cognized in advance: what-it-is and that-it-is-so.

Ad 1.3 To the third, I reply that it cannot be cognized in advance that something is of some sort or quantity, since understanding something to be of a certain sort is understanding a quality to inhere in it, and we have scientific knowledge that a quality inheres in something through demonstration. But nothing known through demonstration can be cognized in advance, and so we cannot cognize in advance that something is of some quality or quantity.

---

[3] The two advance cognitions about something, then, are what is signified by the name (*quid nominis*) and that it is (*quia est*). The replies make clear that these two questions are precognized about all three of the items initially mentioned in the question: the subject, the attribute, and the axiom.

⟨Questio 4⟩

Consequenter queritur de secundo

1 Et arguitur quod sic, quia quod quid est rei significatur nomine rei, sicut quod quid est hominis significatur nomine hominis; sed definitio hominis dicit quod quid est hominis; ergo dicit quid significatur nomine hominis, et per consequens quid significat nomen; ergo eadem est definitio dicens quid est rei et nominis.

2 Oppositum arguitur sic: definitio dicens quid est nominis et quid est rei non est eadem, quia res et nomen non sunt idem; ergo nominis definitio, dicens quid significat nomen, non est definitio dicens quid est res.

3 Ad istam questionem, dico quod non est eadem definitio dicens quid est nominis et dicens quid est rei, quia per Philosophum[a] secundo huius questio quid est presupponit questionem si est. Et ideo dicit Avicenna[b] in *Metaphysica* sua: qui dicit aliquid esse rem et habere quidditatem, et dicat eam non esse, non est de universitate eorum qui aliquid cognoscunt. Ex hoc accipio quod solius entis est definitio significans quid est res, sed definitio indicans quid significat nomen est entis et non entis.

Item, dicit Themistius[c] super istum passum quod si aliquis precipiat puero ire ad ⟨s⟩tabulum ubi sunt equi et asini, et quod adducat equum, adducet equum, et hoc non faceret nisi sciret quid significaretur nomine equi, et tamen ignorat quid est equum secundum substantiam suam.

Item, possumus scire quid significatur nomine passionis ante demonstrationem, sed [scire] quid est passio non possumus scire ante demonstrationem, cum illud sciatur ex inherentia passionis ad subiectum.

---

[a] Aristoteles, *AnPo*, II, 1, 89 b 32–35   [b] Avicenna latinus, *De Philosophia prima vel scientia divina*, I, 5 (ed. Van Riet, pp. 39–40)   [c] Themistius, *Analyticorum posteriorium paraphrasis*, 1 (ed. O'Donnell, pp. 246–47)

---

8 ergo] sed M   12 questionem] questio M   23 illud] illa M

Question 4.

Next we inquire about the second question whether the definitions indicating what a thing is and what the name signifies are one and the same.[4]

1 It is argued that they are, since what a thing is is signified by the name of the thing, for instance, what a human being is is signified by the name 'human being'. But the definition of human being indicates what a human being is, therefore it indicates what is signified by the name 'human being', and consequently it indicates what the name signifies. Therefore the definition indicating what a thing is and the definition of the name are the same.

2 An argument for the opposite side is this: a definition indicating what the name is of and a definition indicating what the thing is are not the same, since a thing and a name are not the same; therefore the definition of the name indicating what the name signifies is not a definition indicating what the thing is.

3 In response to this question I maintain that the definition indicating what the name is of and the definition indicating what the thing is are not the same. For the Philosopher, in *Posterior Analytics* Book II, says that the question of what-it-is presupposes the question of if-it-is, and therefore Avicenna says in his *Metaphysics* that whoever says something is a thing and has a quiddity and also says it does not exist does not belong to the company of those who know anything. From this, I accept that a definition signifying what a thing is is only of a being, but a definition indicating what a name signifies is of beings and non-beings alike.

Furthermore, Themistius says, commenting on this passage, that if anyone orders a boy to go to a stable where there are both horses and donkeys, and to lead out a horse, he will lead out a horse, and he would not have done this if he had not known what is signified by the name 'horse', and yet he does not know what a horse is according to its substance.

Furthermore, we can know what is signified by the name of an attribute before demonstration, but we cannot know what the attribute is before demonstration, since that is known from the attribute's inherence in the subject.

---

[4] Cf. q. 3, l. 1–4.

Item, dicit Avicenna[a] quod omnium illorum quorum est definitio significans quid est nomen est definitio significans quid est esse rei, vel secundum esse quod habet in anima vel secundum esse in re extra. Unde non cuiuslibet est definitio dicens quid est esse rei extra cuius est definitio dicens quid significatur per nomen; possumus enim scire quid significatur nomine vacui, scire tamen quid vacuum sit non possumus, cum non sit. Sed quod Philosophus[b] dicit in IV *Metaphysice*, quod possumus scire quid significatur per nomen non sciendo quid res est, dicit Avicenna[c] quod omne quod habet definitionem habet esse in re extra vel in anima.

Ad 1 Ad rationem, cum dicitur 'quod quid est rei significatur nomine rei', dico quod verum est. Et tu dicis: ergo definitio dicens quid est rei est definitio dicens quid significatur nomine rei. Dico quod non sequitur, immo sequitur quod illa definitio dicet quid est secundum suam substantiam quod nomine rei significatur; dicere autem hoc non est dicere quid nomen significat.

⟨Questio 5⟩

Consequenter queritur circa partem illam: *scire opinamur* **etc. (71 b sq.), primo utrum scientia possit generari in nobis per doctrinam.**

1.1 Et videtur quod non, quia quod non generatur in nobis, impossibile est generari in nobis per doctrinam; sed scientia non generatur in nobis; ergo etc. Probatio minoris: illud enim quod advenit alicui per sedationem et quietationem non generatur in eo, quia sedatio et quietatio opponitur generationi, que est transmutatio. Sed anime advenit scientia per sedationem et quietationem; vult enim Philosophus[d] quod anima in sedando et quiescendo fit sciens et prudens; quare scientia non generatur in anima nostra, ergo etc.

1.2 Item, si doctor generet in nobis scientiam, hoc esset per suam scientiam, sed per suam scientiam non potest, quia tunc sua scientia haberet

---

[a] Avicenna latinus, *De philosophia prima vel scientia divina*, I, 5 (ed. Van Riet, p. 36)   [b] Aristoteles, *Met.*, IV, 4, 1006 a 19–32   [c] Avicenna latinus, *De philosophia prima vel scientia divina*, I, 5 (ed. Van Riet, p. 36)   [d] Aristoteles, *Phys.*, VII, 3, 247 b 10–12

---

24 item] unde *M*

Furthermore, Avicenna says that of each of those things of which there is a definition signifying what the name is of, there is also a definition signifying what-it-is for the thing to exist, either according to the being it has in the soul or according to its being in external reality. And hence for each thing of which there is a definition indicating what is signified by the name, there is not also a definition indicating what-it-is for the thing to exist externally. For we can know what is signified by the name 'vacuum', and yet we cannot know what a vacuum is, since there is no vacuum. But as for what the Philosopher says in *Metaphysics* IV, that we can know what is signified by a name, not knowing what the thing is, Avicenna says that everything with a definition has being either in external reality or in the soul.

Ad 1 In reply to the argument, when it is said that what a thing is is signified by the name of the thing, I reply that this is true. Then you say therefore the definition indicating what the thing is is a definition indicating what is signified by the thing's name. I hold that this does not follow, for even though it follows that the definition will indicate what-it-is according to its substance that is signified by the name of the thing, still, to say this is not to say what the name signifies.

Question 5.

Next, with respect to the phrase 'To know, in our opinion', etc., we ask, first, whether knowledge can be generated in us through teaching.

1.1 It seems it cannot, since it is impossible for that to be generated in us through teaching which is not generated in us at all. But knowledge is not generated in us at all. Therefore etc. Proof of the minor premise: That which comes to belong to something through its being calmed and brought to rest is not generated in it, since being calmed and at rest are opposed to generation, which is change. But knowledge comes to the soul through its being calmed and brought to rest, for the Philosopher says that in being calmed and brought to rest the soul comes to be knowing and wise. So knowledge is not generated in our soul. Therefore etc.

1.2 Furthermore, if a teacher generated knowledge in us this would occur through his knowledge. But it cannot be done through his knowledge

virtutem activam; universaliter enim illud mediante quo agens agit virtutem activam habet super illud quod agitur. Sed scientia non habet virtutem activam, cum sit de prima specie qualitatis. Doctor ergo per suam scientiam non potest in aliquo scientiam generare, ergo scientia non potest in aliquo generari per doctrinam.

2 Oppositum arguitur: perfectum est quod potest generare sibi simile in alio; si ergo doctor est perfectus in scientia, poterit generare consimilem habitum in discipulo, et | ita poterit scientiam generari in aliquo per doctrinam.   M 100ra

3 Intelligendum quod opinio Platonis fuit quod scientia esset in anima nostra penitus ab intrinseco, ita quod omnem scientiam et omnem habitum scientie habet in actu a principio sue originis. Et tunc querebatur ab eo: cum habeamus in nobis habitum scientie, quare non consideramus quando volumus? Respondebat quod hoc erat propter unionem ad corpus. Ex tali enim unione inclinatur ad delectationes sensuales, que ut plurimum retrahunt hominem a studio scientie et ab eius actuali consideratione. Unde anima corpori unita omnia cognoscit in universali, sed per conversionem eius ad ideam separatam cognitionem omnium habet in particulari.

Sed ista opinio veritatem non habet. Philosophus[a] enim dicit tertio *De anima* quod intellectus ante addiscere vel invenire nihil est eorum que sunt et est quasi tabula rasa in potentia ad omnes scientias; hoc autem non esset si anima nostra a principio sue originis haberet habitum scientie.

Item, Philosophus[b] secundo *De anima* dicit quod aliquis est in potentia sciens dupliciter: vel quia habet habitum scientie, non tamen actu considerat, sicut geometer qui habet habitum geometrie, non tamen actu considerat, vel quia nec habet habitum nec actu considerat, sicut pue⟨r⟩ qui nihil scit de geometria. Sed manifestum est quod, si scientia esset in nobis sicut Plato posuit, puer non diceretur potentia sciens secundum predictum modum; quare hec opinio directe est contra Philosophum.

Est igitur dicendum quod scientia potest generari in nobis ab extrinseco et per doctrinam. Ad cuius evidentiam considerandum est quod quedam fiunt ab arte, ut domus, quedam a natura, ut plante et animalia, quedam

---

[a] Aristoteles, *De an.*, III, 4, 429 b 32–430 a 2   [b] Aristoteles, *De an.*, II, 5, 417 a 22–b 1

---

28 a] ab M   32 veritatem] v^{i}tutem M *ut v.*

since then his knowledge would have an active power, for in every case what an agent uses to act has an active power over that which is acted upon. But knowledge does not have active power because it belongs to the first species of quality. Therefore a teacher cannot generate knowledge in anyone through his knowledge; therefore knowledge cannot be generated in anyone through teaching.

2 To the contrary: That is perfect that can generate in another something similar to itself. If then a teacher is perfect in knowledge he will be able to generate a similar disposition in a student, and so will be able to generate knowledge in someone through teaching.

3 It should be understood that the opinion of Plato was that knowledge is in our soul's innermost part intrinsically, so that it has all knowledge and every disposition of knowledge actually from its initial origin. And when he was asked, 'since we have within us disposition knowledge, why do we not bring it to mind when we wish to?' he responded that this was due to our union with the body, for from such a union one is inclined to sensory pleasures that strongly draw a man away from interest in knowledge and from actually bringing it to mind. Hence a soul united to a body cognizes everything universally, but through its turning to the separated Idea it has cognition of all things in particular.

But this opinion lacks truth, for the Philosopher says in *De Anima* III that the intellect, before it learns or discovers, is none of the things that are, and is, as it were, a blank tablet, in potentiality toward all knowledge. But this would not be if our soul had a disposition toward knowledge from its initial origin.

Furthermore, the Philosopher, in *De Anima* II, says that someone knows potentially in two ways, first because he has the disposition toward knowledge, but does not actually bring it to mind, as with the geometer who has the disposition toward geometry but does not actually bring it to mind; second because he neither has the disposition nor actually brings it to mind, as with a boy who knows nothing of geometry. But it is obvious that if there were knowledge in us as Plato thinks then the boy would not be said to know potentially in the aforesaid way, so that this opinion is directly opposed to the Philosopher.

It should be said, then, that knowledge can be generated in us by something extrinsic and through teaching. To make this evident it should be noticed that some things arise from art, for instance a house, others

ab utroque, ut sanitas. Partim enim fit a natura et partim ab arte, a natura quidem tamquam ab agente principali, ab arte vero tamquam ab agente instrumentali. Natura enim in agendo sanitatem in animali est agens principale in dirigendo humores crudos et expellendo superfluos; medicus autem, qui est agens per artem, est agens instrumentale et coadiuvens, iuvat enim naturam per potionem, per balnea et per amplastra, et si virtus naturalis interior deficeret, medicus numquam sanitatem induceret. Recte sicut est in generatione sanitatis, simili modo est de generatione scientie in nobis, nam ipsa generatur in nobis a doctore tamquam agente coadiuvante et removente prohibens, sed generatur ab aliquo principio per se noto tamquam ab agente intrinseco et principali. Verbi gratia, scientia istius conclusionis ⟨habere tres⟩ generatur in discipulo, si sciat illud per se notum quod angulus sit equalis duobus angulis intrinsecis sibi oppositis, et si sciat illud, tunc doctor applicat ad conclusionem et inducit scientiam removendo prohibens. Si autem discipulus ⟨non sciat⟩ quod angulus extrinsecus etc., tunc oportet quod doctor continue resolvat usque ad rationem entis et terminorum; hanc enim nullus intellectum habens potest ignorare. Unde si doctor debet causare scientiam in discipulo, oportet quod discipulus supponat aliquid per se notum in ratione. Et ad hoc attendens ille Lincolniensis[a] dixit quod, si vere loqui velimus, nec qui exterius sonat docet, nec littera scripture vis[u]a exterius docet, sed solum movent hec duo et excitant, sed verus doctor est qui mentem illuminat et veritatem ostendit. Et quid est hoc? Certe ratio primorum principiorum que, secundum quod dicit Albertus,[b] sunt prima lumina illuminantia intellectum possibilem.

Ad 1.1 Ad rationem primam, cum dicitur 'impossibile est scientiam generari in nobis, ergo et per doctrinam', dico quod consequentia est necessaria, sed antecedens nego. Et cum probatur 'quod advenit alicui per sedationem et quietationem non generatur', verum est de generatione secundum quod est cum adiectione contrarii et corruptione simpliciter, sed non est verum de generatione secundum quod generatio notat adeptionem alicuius perfectionis sine transmutatione que debetur sibi per se, sicut cum

---

[a] Robertus Lincolniensis, *In Posteriorum analyticorum libros*, I, 1 (ed. Rossi, p. 94) [b] Albertus Magnus, *In Analytica Posteriora*, I, I, 3 (ed. Borgnet, p. 8).

53 recte ... sanitatis] *post* instrumentali M   55 principio] paralogismo M   66 scripture] ubi *add. sed del.* M

arise from nature, as do plants and other such things, and others from both art and nature, for instance health. For knowledge arises partly from art and partly from nature, from nature as from a principal agent, but from art as from an instrumental agent. For nature is the principal agent in causing health in animals, regulating the undigested humors, and expelling superfluous humors. The doctor, who is the agent by art, is an instrumental and assisting agent. For he aids nature through potions, through baths, and through plasters, and if the internal natural power were to be deficient then the doctor would not bring about health. And just as it happens in the generation of health, so it is in a similar way for the generation of knowledge in us. For it is generated in us by a teacher as an assisting agent who removes hindrances, whereas it is generated in another way by a principle known *per se*, in the way appropriate to an intrinsic and principal agent. For example, the knowledge of the conclusion that a triangle has three angles equal to two right angles is generated in a student if he knows the *per se* knowable proposition that an exterior angle of a triangle is equal to the sum of the two interior angles opposite it. And if he knows that, then the teacher applies it to the conclusion and produces knowledge in him by removing what prevents it. But if the student does not know that the exterior angle etc., then it is necessary that the teacher resolve this progressively until he reaches the nature of the thing in question and the terms predicated of it—for no one with an intellect can be ignorant of these. Hence, if the teacher is to cause knowledge in the student it is necessary that the student attain something knowable *per se* in its nature. And with this in mind Grosseteste said that if we wish to speak truly neither does he teach who makes sounds externally, nor does the externally visible letter teach in writing, but these two only move and excite, and the true teacher is what illuminates the mind and shows forth the truth. And what is this? Surely it is the nature of the first principles which, according to what Albert says, are first lights illuminating the possible intellect.

Ad 1.1 In response to the first argument, when it is said that it is impossible to generate knowledge in us, therefore not even through teaching, I reply that the entailment is necessary but I deny the antecedent. And when it is proved that what comes to one through being calm and brought to rest is not generated, this is true of that generation which comes with the arrival of a contrary and destruction without qualification, but it is not true of that generation which marks the attainment of a perfection

aer sit vel generetur luminosus, et isto modo scientia potest generari in anima per sedationem et quietationem.

Ad 1.2 Ad aliam, dico quod scientia bene habet talem virtutem activam qualis sufficit ad agens coadiuvans. Vel potest dici quod, licet scientia per se non sit qualitas activa, tamen per sermonem significativum generatur et per ipsum agit in discipulum; formatur enim sermo a doctore et recipitur in discipulo.

⟨Questio 6⟩

Consequenter queritur utrum sit nobis possibile scire aliquid.

1.1 Et videtur quod non, quoniam nos non scimus aliquam veritatem nisi sciamus causam simpliciter primam; sed causam simpliciter primam scire non possumus; ergo penitus nullam veritatem scimus. Maior declaratur, quoniam causa prima est causa cuiuslibet veritatis; sed ad cognitionem effectus requiritur cognitio cause; quare ad cognitionem cuiuslibet veritatis requiritur cognitio | cause prime. Minor etiam manifesta est per auctorem *De causis*,[a] qui dicit quod causa prima superior est omni narratione et lingue deficiunt a narratione eius, narrationem autem accipit pro cognitione; causa igitur prima excedit cognitionem nostri intellectus; quare etc.

1.2 Item, nihil est quod non habeat iudices contrarios; de primo enim principio fuerunt aliqui iudices et opponentes contrarie. Qua ratione igitur est standum iudicio unius, eadem ratione est standum iudicio alterius. Sed non existente iudicio cui magis credendum sit, non est possibile determinate aliquid scire. Ergo manifestum est quod scientia cuiuslibet rei est nobis impossibilis.

2 Oppositum arguitur: appetitus naturalis non est ad impossibile, quia esset frustra; sed omnes homines naturaliter appetunt adquirere aliquam scientiam; ergo illud est possibile cuilibet homini, quare possibile est nobis aliquid scire. Et hec est ratio Commentatoris[b] II *Metaphysice*.

---

[a] *Liber de causis* V (VI) 57 (ed. Pattin, p. 147)   [b] Averroes, *Commentarium magnum in Metaphysicam Aristotelis*, II, I, comm. 1 (1562, f. 28K)

---

78 sit]   sic M   3–5 causam ... quoniam]   *iter.* M   17 quia]   quare M

without change which it owes to itself *per se*, as when air is, or is generated to be, transparent. And in this way knowledge can be generated in the soul through its being calmed and brought to rest.

Ad 1.2 In response to the other I reply that knowledge certainly does have such an active power as suffices for acting as an assistant. Or else, it can be said that although knowledge by itself is not an active quality, nevertheless it is generated through signifying expressions and it acts on students through such expressions. For the expression is formed by the teacher and received in the student.

Question 6.

Next it is asked whether it is possible for us to know anything.

1.1 It seems that it is not, since we do not know any truth unless we know the cause which is first without qualification, but we cannot know the cause which is first without qualification; therefore we know no truth at all. The major premise is explained as follows: The first cause is the cause of every truth, but cognition of the cause is required for cognition of the effect, whence the cognition of the first cause is required for the cognition of every truth. The minor premise is also obvious from the author of *De causis*, who says that the first cause is above every explanatory account (*narratio*), and the tongue lacks an explanatory account of it. Now, he uses 'explanatory account' for 'cognition'. The first cause, then, exceeds the cognition of our intellect, therefore etc.

1.2 Furthermore, there is nothing about which people do not reach contrary judgments. For even concerning the first principle there have been opponents who judged it in contrary ways. So for whatever reason we should adhere to the judgment of one, for the same reason we should adhere to the judgment of the other. But if there is no judgment that is to be believed more, then it is not possible to know anything determinately. Therefore it is obvious that knowledge of anything is impossible for us.

2 On the other hand: It is argued that a natural desire is not directed to what is impossible, since that would be in vain. But all human beings naturally desire to acquire some knowledge, therefore it is possible for each human being to do so, so that it is possible for us to know something. And this is the argument of the Commentator, in the second book of the *Metaphysics*.

3 Dicendum ad questionem quod scientia alicuius rei est possibilis nobis, quod apparet ex duobus. Primo, quia si cause que habent inducere aliquem effectum sint possibiles, tunc effectus erit possibilis. Ponere enim aliquam causam cuius effectus sit impossibilis est removere ab ea rationem cause, quia causa est ad cuius esse sequitur aliud. Sed cause ex quibus est generatio ⟨scientie⟩ sunt possibiles, non solum simpliciter, sed etiam nobis. Ad generationem enim scientie concurrunt duo principia: principium passivum, quod est intellectus possibilis, et principium activum, quod est phantasma et intellectus agens. Unde intellectus agens ad presentiam phantasmatis in virtute primorum principiorum, que sunt prima lumina illuminantia intellectum possibilem, ut dicit Albertus,[a] ⟨est agens⟩ introductionis et generationis scientie. Et ideo cum omnia ista possibile sit in nobis reperire, immo etiam necessarium est, manifestum est quod possibile est scientiam in nobis generari.

Et hoc etiam apparet ex alio, nam moveri et motum esse unius rationis sunt quantum ad hoc quod, si unum est possibile, et reliquum erit possibile. Sed nos videmus quod possibile est homines moveri ad scientiam; immo etiam moventur querentes cognitionem posteriorum per priora, aliqui quidem ex inventione, aliqui ex aliorum instructione. Et ideo manifestum est quod hominibus inest potestas ad comprehensionem scientie, ita ut non solum fuit ad moveri ad scientiam, immo etiam ad motum esse.

Secundum autem Commentatorem[b] duo sunt signa ex quibus perpenditur si quis rectam scientiam habet de aliquo: unum est si opinionem quam habet possit reducere in sensibilia et per se nota, quia experimentum sermonum verorum est ut conveniant rebus sensatis; aliud est si ipse possit reducere quemcumque alium ad hoc quod non possit sue opinioni contradicere per rationem scientificam. Manifestum est tunc quod philosophus talem opinionem habens dispositus est ad scientiam.

Sic ergo per rationem et etiam per signa manifestum est quod scientia est nobis possibilis, in quo manifeste apparet error Academicorum qui dicebant nos omnia ignorare et nullius rei cognitionem habere.

Ad 1.1 Ad rationem primam, cum arguitur 'nos non scimus etc.', verum est. Dico quod prima causa veritatis potest esse duplex, vel que est prima

---

[a] Albertus Magnus, *In Analytica posteriora*, I, I, 3 (ed. Borgnet, p. 8) [b] Averroes, *Commentarium magnum in Physicam Aristotelis*, VIII, 8, comm. 22 (1562, f. 357BF)

---

38 aliqui] aliquid *ante corr.* M   41 ad motum] a moto M

3 It should be said in response to the question that knowledge of some thing is possible for us, which becomes apparent in two ways: In the first place, if the causes that must lead to some effect are possible, then the effect will be possible. For to posit a cause the effect of which is impossible is to remove from it the nature of a cause, since a cause is that upon the being of which another follows. But the causes from which there is generation of knowledge are possible, not merely without qualification, but for us. For two principles concur in the generation of knowledge, a passive principle, the possible intellect, and an active principle, the phantasm and the agent intellect. Hence the agent intellect, in the presence of a phantasm and through the power of first principles—which as Albert says are the first lights illuminating the possible intellect—is the agent for the introduction and generation of knowledge. Therefore, since it is possible for all this to be found in us—indeed, it is found in us necessarily—it is obvious that it is possible for knowledge to be generated in us.

And this is also apparent from another argument, for to be moved and to have been moved have the same account in this regard, that if one is possible then the other will be possible, too. But we see that it is possible for a human being to be moved towards knowledge, and especially those who seek cognition of the posterior through the prior are so moved: some through discovery and some through the instruction of others. And therefore it is obvious that a power for the comprehension of knowledge is in human beings, in such a way that not only was this power there for being moved toward knowledge, but also for having been so moved.

But according to the Commentator there are two signs from which are discerned one's rightly having knowledge about anything. One is that the opinion he has can be brought back to sensibles and things known in themselves, for the test of true statements (*sermonum*) is that they agree with things that are sensed. The other is whether he can bring anyone else back to a state of being unable to contradict his opinion through a knowledge-producing scientific argument. It is plain, then, that a philosopher holding such an opinion has the disposition for knowledge.

Thus, through argument and through these signs, it is plain that knowledge is possible for us; and in this the error is plain of the Academics, who said we are ignorant of all things, and have cognition of nothing.

Ad 1.1 In response to the first argument, when it is argued 'we do not know' etc., this is true. I hold that the first cause of truth can be taken in

in essendo vel que est prima solum in cognoscendo. Causa prima in essendo cuiuslibet veritatis est ipsum primum principium quod est summe ens et summe verum, causa prima in cognoscendo, saltem quoad nos, sunt prima principia [in] que, sicut dicit Albertus,[a] sunt quedam instrumenta per que intellectus noster educitur ut fiat actu sciens. Quod ergo dicitur quod non scimus nisi sciamus causam primam veritatis, dicendum quod verum est de causa prima in cognoscendo, sed non de causa prima in essendo.

Sed tu dices: causa prima in essendo est causa cuiuslibet veritatis; sed ad cognitionem completam cuiuslibet effectus requiritur cognitio cause; ergo et ad cognitionem cuiuslibet veritatis. Dico quod duplex est cognitio completa, scilicet cognitio completa et perfecta simpliciter et cognitio completa in genere. Tunc dico quod ad cognitionem completam in genere cuiuslibet effectus non requiritur cognitio cause simpliciter prime, sed verum est quod cognitio cause prime requiritur ad cognitionem simpliciter completam de effectu, tamen hoc modo nihil cognoscitur a nobis cognitione completa, sed primo modo solum.

Ad 1.2 Ad aliud, cum dicitur 'qua ratione credendum est uni, et alteri eadem ratione habenti opinionem contrariam', dicendum quod non equaliter credendum est iudicio cuiuslibet, sed magis credendum est iudicio sapientis quam ignorantis et vigilantis quam dormientis.

Et tu dicis: hoc habet iudices contrarios, quod iste sit sciens vel ignorans. Dicendum quod qui bene dispositus est secundum mentem statim percipit an iste sit sciens vel ignorans; signa autem ex quibus hoc perpendi potest superius posita sunt.

⟨Questio 7⟩

Consequenter queritur utrum | passionis et aliorum accidentium sit definitio et quod quid est. M 100va

1.1 Et arguitur quod non, quia quod quid est est quod est hoc aliquid et unum; sed accidens non est huiusmodi, sed magis quale vel quantum; ideo etc.

---

[a] Albertus Magnus, *In Analytica posteriora*, I, I, 3 (ed. Borgnet, p. 8)

---

4 quale] quid *add. sed exp.* M

two ways, either as what is first in being or what is first in cognizing. The first cause in being of every truth is that first principle that is the highest being and the most true. The first cause in cognizing, at least as far as we are concerned, are the first principles which, as Albert says, are certain lights through which our intellect brings it about that we are knowers in act. Therefore, because it is said that we do not know unless we know the first cause of truth, it must be maintained that it is true of the cause that is first in cognizing, but not of the cause that is first in being.

But you will say the first cause in being is the cause of each truth, but for the complete cognition of each effect a cognition of the cause is required, therefore also for the cognition of each truth. I reply that a complete cognition is of two sorts, a cognition complete and perfect without qualification, and a cognition complete in its genus. Then I maintain that it is not required for a cognition of an effect which is complete in its genus that there be a cognition of the cause that is first without qualification. But it is true that a cognition of the first cause is required in a cognition complete without qualification of the effect, although nothing is cognized by us through a complete cognition in this way, but only in the first way.

Ad 1.2 In response to the other argument, when it is argued that whatever reason we have for believing one, we have the same reason to believe the other, who has a contrary opinion, it should be replied that not every judgment is equally to be believed, but the judgment of the wise is more to be believed than that of the ignorant, and that of the waking more than the judgment of the sleeper.

You will answer that there are those who reach contrary judgments regarding whether this one is knowing or ignorant. It should be replied that whoever is well disposed in his mind at once perceives whether this one is knowing or ignorant. The signs from which this may be discerned have been listed above.

Question 7.

Next it is asked whether there is a definition and essence (*quod quid est*) of an attribute and other accidents.

1.1 It is argued that there is not, since an essence is what this something, this unity, is, but an accident is not like this, but is rather this sort or so many; therefore etc.

1.2 Item, quod quid est vult esse aliquid absolutum; sed accidens non est aliquid absolutum; ergo etc. Maior patet, quia vult esse absolutum ab aliis. Minor etiam patet, quia accidentia non definiuntur sine substantia.

2 Oppositum arguitur sic: quod quid est potest esse cuiuslibet comprehensi ab intellectu; sed accidentia possunt comprehendi ab intellectu; ergo etc.

3 Intelligendum est quod accidentium est quod quid est, quia sicut se habet aliquid ad esse, ita ad quod quid est, quia ens dicit illud absolute quod quod quid est dicit in comparatione ad definitionem; sed ipsorum accidentium est esse per se, quia esse per se dividitur in decem praedicamenta. Sed ipsorum accidentium non est quod quid est sicut substantiarum, quia substantiarum est esse simpliciter et primo sine comparatione ad aliquod extrinsecum, accidentium autem est esse secundum quid; accidens enim non dicitur ens nisi quia entis. In definitione ergo et quod quid est accidentis cadit subiectum, et ita illorum non est quod quid est simpliciter et primo, sed secundum quid et secundum accidens, ut includendo ipsum subiectum. Unde Commentator[a] super XI *Metaphysice* dicit quod negare accidentibus quod quid est est negare [ab] eis quod eis competit, attribuere autem accidentibus quod quid est simpliciter est attribuere eis quod non est eis attribuendum, et ideo tenenda est via media, quod habent quod quid est per posterius.

Sed cum accidentia habeant quod quid est et definitionem, quod istorum habent verius? Dico quod verius habent quod quid est quam definitionem significantem quod quid est, quia quod quid est habent sine respectu ad substantiam extrinsecam, quia quod quid est accidentium non est quod quid est substantie, sed definitiones habent per respectum ad substantiam extrinsecam, inde est quod definitio accidentium est per additamentum.

Et per hoc est responsio ad rationes.

Ad 1.1 Ad primam, dico quod accidentia non sunt simplicia omnino, et ideo non competit eis definitio per se et primo.

---

[a] *Locum non invenimus*; cf. Averroes, *Epitome* in *Aristotelis Metaphysicorum libri XIIII cum Averrois Cordubensis in eosdem commentariis et epitome*, XI, summa 3, caput 2 (1562, f. 282E *marg.*)

---

14 ipsorum] definitionem *add. sed exp.* M   24 autem] et M   25 via] una M   28 habent[2]] haberent M *ut v.*   32 per] experimentum *add. sed exp.* M

BOOK I 29

1.2 Furthermore, essence implies an absolute being, but an accident is not something absolute; therefore etc. The major premise is obvious since essence implies being absolute apart from others. The minor is also clear, since accidents are not defined without a substance.

2 The opposite view is argued thus: There can be an essence of each thing that is grasped by the intellect, but accidents can be grasped by the intellect; therefore etc.

3 It should be understood that there is an essence of accidents, since as something stands to being, so it stands to essence; for 'being' indicates that absolutely which the essence indicates with reference to a definition. But there is being *per se* of accidents themselves, since being *per se* is divided into ten categories. But there is not an essence of accidents themselves in the same way as there is of substances, for being belongs to substances without qualification and primarily, without reference to anything extrinsic, but accidents have being only in a certain respect. For an accident is not called a being except because it is of a being. Therefore the subject falls within the definition and essence of an accident, and so there is no essence of these without qualification and primarily, but only in a certain respect and as an accident, by including its subject. Whence the Commentator, on *Metaphysics* XI, says that to deny an essence to accidents is to deny of them what coincides with them, whereas to attribute to accidents an essence without qualification is to attribute to them what must not be attributed to them. And so a middle ground must be maintained: that they have an essence in a secondary sense.

But although accidents have an essence and definition, which of these do they have more truly? I reply that they have an essence more truly than a definition signifying an essence, for they have an essence without respect to an extrinsic substance, since the essence of accidents is not the essence of a substance, but they have a definition with respect to an extrinsic substance. Hence the definition of accidents occurs through something added.

And through this there is a response to the arguments.

Ad 1.1 To the first I reply that accidents are not wholly simple and therefore a primary and *per se* definition does not agree with them.

Ad 1.2 Ad aliam, dico quod quod quid est absolutum dupliciter potest esse: vel quia ad quod quid est non pertinet aliquid extrinsecum, aut ita quod non attenditur ad aliquid extrinsecum; primo modo est quod quid est accidentium absolutum, secundo modo non.

⟨Questio 8⟩

Consequenter queritur utrum addiscens aliquam conclusionem prius scivit eam.

1.1 Et arguitur quod sic, quia addiscens conclusionem aut prius scivit eam aut non. Si sic habetur propositum. Si non, ergo si obviet ei, non magis apprehendet de ea quod est illa cuius cognitionem querit quam alia, et ita nihil determinate querit, et hoc est inconveniens.

1.2 Item, intellectus non accipit aliquid de novo nisi per aliquam alterationem factam in eo; sed circa partem intellectivam anime nulla fit alteratio; ergo etc. Minor patet per Philosophum[a] VII *Physicorum*.

2 Oppositum arguitur, quia si addiscens sciret illam conclusionem prius, addiscere non esset nisi reminisci.

3 Ad istam questionem, dico quod addiscens aliquam conclusionem scivit eam prius in universali. Secundum enim quod dicit Commentator[b] III *De Anima*, sicut se habet materia ad formas particulares quod primo recipit formas corporum simplicium et postea corporum compositorum, sic se habet intellectus ad formas universales quod primo comprehendit principia et confusa magis et postea applicat ad conclusionem. Unde addiscens aliquam conclusionem prius scivit eam in universali et in suis principiis (secundum enim Lincolniensem[c] principia habent virtutem activam et universalem super conclusionem), tamen non prescivit eam in particulari. Unde addiscens conclusionem scit eam prius in universali et in virtute, tamen ignorat in particulari.

---

[a] Aristoteles, *Phys.*, VII, 3, 247 b 1–248 a 9   [b] Averroes, *Commentarium magnum in Aristotelis De anima libros*, III, 6, comm. 21 (ed. Crawford, p. 455)   [c] Robertus Lincolniensis, *In Posteriorum analyticorum libros*, I, 1 (ed. Rossi, p. 97)

---

36 est] est *add. sed exp.* M   3 quod] non *add. sed del.* M   7 aliquid] alio M   19 virtutem] unitatem M

Ad 1.2 To the other I reply that an essence can be absolute in two ways, in one way because something extrinsic does not belong to it; in the other way because it is not directed toward something extrinsic. The essence of accidents is absolute in the first way, but not in the second.

Question 8.

Next it is asked whether one who learns a conclusion knew it beforehand.

1.1 It is argued that he did, since one who learns a conclusion either knew it beforehand or did not. If he did, we have what is proposed. If he did not then if he does obtain it he will not recognize that it is that which he seeks to cognize rather than something else, and so he seeks no definite thing, and this is absurd.

1.2 Furthermore, the intellect does not receive anything anew unless through some alteration made in it. But there is no alteration with reference to the intellectual part of the soul. Therefore etc. The minor premise is clear from the Philosopher in *Physics* VII.

2 On the other hand, it is argued that if one who is learning knows the conclusion beforehand, this is not learning, but remembering.

3 In response to this question, I maintain that one who learns a conclusion knew it beforehand in the universal case. For according to what the Commentator says on *De anima* III, just as matter is related to particular forms in such a way that it first receives the forms of simple bodies and afterwards those of complex bodies, so the intellect is related to universal forms in such a way that it first grasps principles and things that are more confused and afterwards applies them to the conclusion. Whence one who learns a conclusion knew it beforehand in the universal case, and in its principles (for according to Grosseteste principles have an active and universal power over their conclusion). But one who knows a conclusion did not know it beforehand in the particular case. Therefore, one who learns a conclusion knows it beforehand in the universal case, and virtually, but is ignorant of it in the particular case.

Unde ad rationes.

Ad 1.1 Ad primam, dico quod addiscens conclusionem scivit eam prius in universali. Sed tu probas quod prius scivit eam in particulari, quia aliter non cognosceretur, cum occurret, sicut ponit Themistius[a] qui ponit exemplum de servo fugitivo. Unde intelligendum quod addiscentem conclusionem aliquam oportet prius habere cognitionem aliquam magis determinatam ad eam quam ad aliam, verbi gratia, paterfamilias querens servum fugitivum, nisi magis haberet cognitionem determinatam de eo quam de alio, numquam apprehenderet. Eodem modo est de addiscente aliquam conclusionem. Unde cognitio in universali habetur in demonstratione per maiorem propositionem, cognitio magis determinata explicatur per minorem propositionem, et ideo in demonstratione indigemus duabus premissis.

Sed tu dices: Philosophus dicit quod addiscens aliquam conclusionem prius solum scivit eam in universali. Dico quod verum est quod solum scivit eam in universali prius tempore, nihilominus tamen oportet quod habeat aliquam cognitionem magis determinatam prius natura. Unde, quia ista cognitio ultima habetur per minorem, minor scitur prius natura; prior cognitio habetur [prius] per maiorem, ideo maior scitur prius tempore.

Ad primam rationem patet.

Ad 1.2 Ad aliam, dico quod alteratio potest intelligi dupliciter: aut adeptio alicuius perfectionis sine aliqua | transmutatione facta a contrario, aut transmutatio que fit mediantibus qualitatibus activis et passivis. Unde maior vera est de alteratione primo modo dicta. Ad minorem, dico quod circa partem intellectivam anime non accidit alteratio que est mediantibus qualitatibus activis et passivis, numquam enim fit intellectus calidus vel frigidus, humidus vel siccus, tamen in eo est alteratio que est adeptio perfectionis, et hoc non est inconveniens.

M 100vb

---

[a] Themistius, *Analyticorum posteriorium paraphrasis*, 1 (ed. O'Donnell, pp. 246–47)

---

28 cognitionem] conclusionem M   37 solum[1]] *post* scivit M   39 cognitionem] conclusionem M   48 accidit] accipit M

From this I reply to the arguments.

Ad 1.1 To the first I reply that one who learns a conclusion knew it beforehand in the universal case. But you will prove that he knew it beforehand in the particular case since otherwise he would not recognize it when he happens on it, as Themistius says, who presents the example of the fugitive slave. Whence it must be understood that it is necessary for one who learns a conclusion to have a cognition beforehand adapted more to that conclusion than to any other. For example, the head of a family seeking a fugitive slave without having a cognition more adapted to pick out that slave than any other will never recognize him. The same applies to one who learns a conclusion. Whence a cognition in the universal case is possessed in a demonstration through the major premise, and a more determinate cognition is uncovered through the minor premise. And so in a demonstration we need two premises.

But you will reply that the Philosopher says that one who learns a conclusion knew it beforehand only in the universal case. I answer that it is true that, with respect to temporal priority, he knew it only in the universal case, but nevertheless it is necessary that he have a more determinate cognition that is prior by nature. Hence, since this final cognition is had through the minor premise, the minor is known beforehand by nature; the prior cognition is had through the major premise, so the major is known prior in time.

The response to the first argument is clear.

Ad 1.2 I reply to the other that alteration can be understood in two ways: Either as the attainment of a perfection without some change produced by a contrary, or as a change which arises through the mediation of active and passive qualities. Whence the major premise is true of alteration understood in the first way. I reply to the minor premise that an alteration that arises through the mediation of active and passive qualities does not occur in the intellectual part of the soul, since heat, cold, wetness, or dryness never arises in the intellect. But there does arise in it an alteration that is the attainment of a perfection, and this is not absurd.[5]

---

[5] The intellect is capable only of becoming more or less perfectly what it is. This cashes out as: the intellect (and the soul) can change only in virtue of exercising functions essential to its nature. Thus it can come to know or forget. Lacking matter, the intellect does not undergo alteration in the way characteristic of matter, i.e., via changes in active and passive qualities associated with the four elements. Hence, the intellect does not change with respect to characteristics that are irrelevant to its nature, but only with respect to the degree of its perfection.

⟨Questio 9⟩

Consequenter queritur utrum ista propositio sit vera 'unumquodque propter quod, et illud magis'.

1.1 Et arguitur quod non, quia aliquis est calidus propter motum, et tamen motus non est magis calidus.

1.2 Item, ferrum ignitum est calidum propter ignem, et tamen ignis non est magis calidus.

1.3 Item, Sortes est homo propter Platonem generantem, et tamen non Plato magis homo.

1.4 Item, homo est ebrius propter vinum, et tamen non vinum magis.

1.5 Item, flamma lucet propter ignem, et tamen non ignis magis. Ergo ista propositio est falsa.

2 Oppositum arguitur sic: causa est nobilius causato, quia secundum Proclum[a] omne productivum alterius est nobilius producto; sed illud propter quod est aliud est causa illius al⟨ter⟩ius; ergo si illud aliud eveniat, illud magis eveniet propter quod illud aliud est, cum sit nobilius illo.

3 Ad evidentiam istius, intelligendum est quod 'propter quod' dicit causalitatem, unde significatur per propositionem quod illud quod est causa alicuius dispositionis in aliquo, ipsum est magis tale. Sed causa duplex est: causa univoca et causa equivoca. Univoca causa est que producit in alio quod est in se ipso, sicut ignis in igniendo, sed equivoca est que causat in aliud quod non est in se ipso formaliter, ut sol causat caliditatem et tamen non est calidus. Unde ista propositio potest intelligi vel in causis univocis vel equivocis. Si in univocis, sic habet veritatem 'unumquodque propter quod, et illud magis': verum est formaliter. Si in equivocis, sic habet veritatem 'unumquodque propter quod, et illud magis': verum est effective.

Item, propositio habet veritatem in causis et causatis essentialiter ordinatis et non ⟨in⟩ accidentaliter ordinatis.

[a] Proclus, *Elementatio theologiae*, 7 (ed. Vansteenkiste, p. 267)

---

15 nobilius] etc. *add. sed exp.* M   23 si] sed M   25 quod] quam M

Question 9.

Next it is asked whether this proposition is true, 'everything because of which ⟨something has some character,⟩ is itself even more ⟨of that character⟩'.

1.1 It is argued that this is not true, since someone is hot because of a motion, yet the motion is not more hot.

1.2 Furthermore, iron placed on a fire is hot because of the fire, and yet the fire is not more hot.

1.3 Furthermore, Socrates is a human being because of Plato's bringing him into being, and yet Plato is not more a human being than is Socrates.

1.4 Furthermore, a human being is drunk because of wine, and wine is not more drunk than the human being is.

1.5 Furthermore, flames give out light because of fire, and yet fire is not more a giver of light than the flames are. Therefore this proposition is false.

2 On the other hand, it is argued as follows: a cause is more noble than what it causes, for according to Proclus everything producing another is nobler than its product; but that because of which another exists is the cause of that other; therefore if that other comes to be, then that because of which the other exists will all the more come to be, since it is nobler than it.

3 To make this clear it should be understood that 'because of which' indicates causality, whence it is signified through the proposition at issue that that which is a cause of some state in something is itself more so. But cause is two-fold, univocal and equivocal cause. A univocal cause is one that produces in another what is in itself, as fire does when it sets something on fire. But an equivocal cause is one that causes in another what is not formally in itself, for instance, the sun causes heat and yet is not hot. Whence this proposition can be understood either with respect to univocal causes or equivocal causes. If understood for univocal causes then 'everything because of which is itself even more' is true formally. But if it is understood for equivocal causes then it is true effectively.

Further, the proposition is true in the case of causes and effects that are essentially ordered, and not in the case of causes and effects that are accidentally ordered.

Item, habet veritatem in causis univoce agentibus que sunt per se sufficientes ad causandum effectum in alio.

Ad 1.1 Unde ad primam rationem, dico quod motus magis calidus est effective. Vel sic: sicut dictum est, propositio habet veritatem in causis essentialiter ordinatis; nunc autem motus non est essentialiter ordinatus ad causandum caliditatem, sed essentialiter est motus actus entis in potentia.

Ad 1.2 Ad aliam, dico, secundum quod dictum est, quod propositio habet veritatem in causis sufficientibus ducere effectum in alio; sed ignis non est sufficiens producere caliditatem sub tanta excellentia in ferro, sed compactio ferri facit ad hoc; et ideo non est propositio vera in talibus.

Ad 1.3 Ad aliam, dico, sicut dictum est, quod propositio habet veritatem in causis essentialiter ordinatis; nunc autem vinum non essentialiter ordinatur ad inebriandum, sed secundum quod vaporosum, secundum Albertum,[a] et hoc accidit.

Ad 1.4 Ad aliam, dico similiter quod propositio habet veritatem in causis essentialiter ordinatis; nunc autem accidit Sorti quod generetur a Platone, quoniam si essentialiter inesset Sorti quod generaretur a Platone, eadem ratione essentiale esset Platonem generari a suo patre, et similiter patri a suo patre, et sic in infinitum, et ita aliquis effectus dependeret ⟨a⟩ causis infinitis essentialiter ordinatis, quod est impossibile; ideo ista propositio non habet veritatem in talibus.

Ad 1.5 Ad aliam, similiter: accidit enim igni quod flammet, quia flamma non inest igni nisi in materia aliena, et ideo in talibus non habet propositio veritatem.

⟨Questio 10⟩

Consequenter queritur utrum scire simpliciter sit per causam.

1.1 Et arguitur quod non, quia si scire sit per causam, tunc quorum non est causa non est scientia; sed primorum principiorum non est causa; quare etc., quod est contra Philosophum.[b]

---

[a] Albertus Magnus, *In Analytica posteriora*, I, II, 5 (ed. Borgnet, p. 31)   [b] Aristoteles, *AnPo*, I, 3, 72 b 19–24; 33, 88 b 36–89 a 1

---

40 essentialiter[1]] sufficienter M

Further, it is true in the case of causes acting univocally that are sufficient by themselves to cause an effect in another.

Ad 1.1 Accordingly, in response to the first argument, I reply that the motion is more hot effectively. Or, as was said, the proposition is true in the case of essentially ordered causes. But motion is not essentially ordered to cause heat; rather, motion is essentially the act of a being in potentiality.

Ad 1.2 To the next I reply, according to what was said, that the proposition is true in the case of causes sufficient to produce the effect in another. But fire is not sufficient in itself to produce heat to such a high degree in iron. Instead, it is the density of the iron that produces this, and therefore the proposition is not true in such cases.

Ad 1.3 To the next I reply that, as was said, the proposition is true in the case of causes that are ordered essentially. But wine is not essentially ordered to making one drunk, but it is ordered to this only inasmuch as it is vaporous, according to Albert, and this is accidental.

Ad 1.4 To the next I reply in the same way that the proposition is true in the case of essentially ordered causes. But it is accidental to Socrates that he is brought into being by Plato, since, if it were essential to Socrates that he was brought into being by Plato, then by the same argument it would be essential to Plato that he was brought into being by his father, and likewise essential that his father came from his father, and so on to infinity. And so some effect would depend on an infinite number of essentially ordered causes, which is impossible. Therefore this proposition is not true in such cases.

Ad 1.5 To the next I reply in the same way, for it is accidental to fire that it produces flames, for flames are not in fire unless the fire is in some foreign matter, and therefore the proposition is not true in such cases.

Question 10.

Next it is asked whether knowing without qualification is through the cause.

1.1 It is argued that it is not, for if knowing is through the cause, then there is no knowledge of those things having no cause. But there is no cause of first principles, therefore etc., which is contrary to the Philosopher.

1.2 Item, ⟨non⟩ omnis demonstratio facit scire per causam, quoniam non demonstratio quia; sed omnis demonstratio facit scire simpliciter; ergo etc.

2 In oppositum est Philosophus.[a]

3 Ad istam questionem, dico quod scire simpliciter est per causam. Sed ad hoc intelligendum quod scire simpliciter non ⟨solum⟩ est causam cognoscere, quia scire proprie loquendo est per causam, cum sit ipsarum conclusionum. Et huius probatio est, quia per idem habet res esse simpliciter et sciri simpliciter; sed per causam habet res esse simpliciter; ergo etc.

Sed intelligendum quod ad scire simpliciter non ⟨solum⟩ requiritur cognitio cause, sed triplex requiritur cognitio: cognitio cause que habetur in cognitione cause actualis, ⟨illa⟩ que habetur in applicatione cause ad effectum, et tertia que habetur per hoc quod impossibile est aliter se habere. Et ideo posuit Aristoteles[b] tres particulas in definitione scire, quod est: 'scire est causam cognoscere' et 'quoniam illius est causa' et 'quoniam impossibile est aliter se habere'; per primam particulam prima cognitio, per secundam secunda, per tertiam tertia.

Sed, cum causa possit dupliciter considerari, vel secundum suam substantiam vel secundum quod causa est, posset aliquis querere utrum ad cognitionem requiratur cognitio cause secundum utrumque modum. Et ad hoc dico quod sic. Nam aliquis potest habere cognitionem cause secundum substantiam cause et ignorare effectum, verbi gratia interpositio terre est causa eclipsis lune, et potest sciri ab aliquo quid est interpositio et si est et ignorare si sit eclipsis lune. Et ad hoc attendens | Philosophus[c]   M 101ra
dixit quod scire est causam cognoscere et quoniam illius est causa: per primum intelligitur quod ad cognitionem effectus requiritur cognitio cause secundum substantiam suam, per secundum quod requiritur cognitio cause secundum quod causa est.

---

[a] Aristoteles, *AnPo*, I, 2, 71 b 10–13   [b] Aristoteles, *AnPo*, I, 2, 71 b 10–13   [c] Aristoteles, *AnPo*, I, 2, 72 a 25–28

---

16 que] non *add. sed exp.* M   18 tertia] certa M   20 causa] causam *ante corr.* M ||
quoniam²] illa *add. sed exp.* M   29 si¹] sic *ante corr.* M || et²] *s. l.* M

1.2 Furthermore, not every demonstration produces knowing through the cause, given that a demonstration that the thing is so does not do so. But every demonstration produces knowing without qualification; therefore etc.

2 Opposed to this is the Philosopher.

3 In response to this question, I reply that knowing without qualification is through the cause. But one should understand that knowing without qualification is not just cognizing the cause, since knowing properly speaking is *through* the cause since the knowledge is of the conclusions. And the proof of this is that a thing has being without qualification and is known without qualification through the same thing, but a thing has being without qualification through a cause. Therefore etc.

But it should be understood that in order to know without qualification what is required is not just a simple cognition of the cause but a three-fold cognition: a cognition of the cause obtained in cognizing the actual cause; a cognition of the cause obtained in the application of the cause to the effect; and, third, what is obtained through its being impossible that it should be otherwise. And therefore Aristotle placed three clauses in the definition of knowing, that is, 'knowing is cognizing the cause, and cognizing that it is the cause of this, and cognizing that it is impossible that it should be otherwise'. The first clause refers to the first cognition, the second refers to the second and the third refers to the third.

But since a cause can be regarded in two ways, either as it is substance or as it is a cause, one can ask whether a cognition of the cause regarded in both of these ways is required for cognition. And I reply to this that both are required, for one could have cognition of the cause regarded as the substance of the cause and fail to know the effect. For example, the interposition of the earth is the cause of a lunar eclipse, and it can be known by someone what an interposition is and if it is, and yet he can be ignorant whether there is an eclipse of the moon. Attending to this, the Philosopher said that knowing is cognizing the cause and that it is the cause of this. It is understood in the first that a cognition of the cause as regards its substance is required for cognition of the effect, and it is understood in the second that a cognition of the cause considered as a cause is required.

Ad rationes.

Ad 1.1 Ad primam, dico quod quorum non est aliqua causa, eorum non est scientia simpliciter; et quia primorum principiorum non est causa aliqua, ideo nec scientia simpliciter.

Sed tu dices: Philosophus dicit quod principia magis sciuntur quam conclusiones. Dico quod hoc est verum de scire in communi, quod est incommutabilis comprehensio veritatis, sed scire [est] proprie est per causam cognoscere, unde intellectus proprie est principiorum.

Ad 1.2 Ad aliam rationem, respondeo interimendo minorem; quedam enim demonstratio est que procedit ex prioribus simpliciter, et illa facit scire simpliciter, quedam que procedit ex prioribus quoad nos, et talis non facit scire simpliciter, et ideo illa minor falsa est que dicit quod omnis demonstratio facit scire simpliciter.

⟨Questio 11⟩

Consequenter queritur utrum ad cognitionem effectus requiritur cognitio omnium causarum.

1.1 Et arguitur quod non, quia definitio substantie est sufficiens cognitio substantie; sed definitio substantie non aggregat in se omnes causas, sed solum materiam et formam; ergo etc.

1.2 Item, medicina est completa cognitio sanitatis, et tamen non considerat omnes causas; secundum enim Philosophus[a] I *Ethicorum* contra Platonem, medicus non considerat primam ideam sanitatis.

1.3 Item, demonstratio mathematica est potissima, secundum Commentatorem[b] II *Metaphysice*; et tamen non considerat omnes; ergo etc.

2 Oppositum arguitur sic: effectus est ex omnibus causis, ergo ex omnibus cognoscitur, cum eadem sint principia essendi et cognoscendi.

---

[a] Aristoteles, *EN*, I, 6, 1097 a 7–14   [b] Averroes, *Commentarium magnum in Metaphysicam Aristotelis*, II, 3, comm. 16 (1562, f. 35K)

---

36 scientia] esse M   9 mathematica] mai^ca M

In response to the arguments.

Ad 1.1 I reply to the first that there is no knowledge without qualification of those things of which there is no cause. So, since there is no cause of first principles, therefore there is no knowledge of them.

But you will say that the Philosopher says that principles are better known than conclusions. I reply that this is true of knowing in the common sense, which is an immutable grasp of the truth. But knowing in the strict sense is cognition through the cause, and so it is understanding, strictly speaking, that concerns principles.

Ad 1.2 To the second argument, I respond by destroying the minor premise, for there is a kind of demonstration that proceeds from what is prior without qualification, and this produces knowing without qualification; and there is another kind of demonstration that proceeds from what is prior for us, and this does not produce knowing without qualification. Therefore the minor is false which says that every demonstration produces knowing without qualification.

Question 11.

Next it is asked whether a cognition of every cause is required for a cognition of the effect.

1.1 It is argued that it is not, since the definition of a substance is a sufficient cognition of the substance, but the definition of a substance does not collect in itself every cause, but only the material and formal cause; therefore etc.

1.2 Furthermore, medicine is the complete cognition of health, and yet it does not consider every cause. For according to the Philosopher in *Ethics* I, speaking against Plato, the doctor does not consider the first idea of health.

1.3 Furthermore, a mathematical demonstration is the highest, according to the Commentator on *Metaphysics* II, and yet it does not consider every cause; therefore etc.

2 On the other hand, it is argued thus: an effect comes from all of its causes, and therefore it is cognized from all of its causes, since the principles from which it has being and from which it is cognized are the same.

3 Ad questionem istam dico quod ⟨ad⟩ cognitionem rei habentis omnes causas oportet cognoscere omnes, intrinsecas et extrinsecas, quia scire est per causam cognoscere; quantum ergo deficit aliquis a cognitione cause, ⟨tantum⟩ deficit a scientia rei.

Item, hoc patet per Aristotelem[a] III *Metaphysice*, qui dicit quod cognitio completa est solutio omnium dubitatorum; sed si sciamus unam causam contingit dubitare de aliis; ergo non est completa cognitio nisi omnes cause cognoscantur.

Item, hoc patet, quoniam non scimus rem quousque sciamus eam resolvere in sua principia omnia; sed hoc non scimus nisi omnes causas sciamus; ergo etc.

Sed intelligendum est quod, cum aliqua scientia consideret rem sub aliqua ratione speciali, sufficit quod consideret causas illius rei secundum quod talis, sicut medicus considerat sanitatem secundum quod sanitas, sed ⟨non⟩ secundum quod ens. Ideo sub illa ratione non oportet illum discurrere ad cognitionem cause prime, cum non sit causa sanitatis secundum quod sanitas, sed secundum quod ens. Similiter Philosophus in libro *De animalibus* non debet considerare materiam primam, cum non sit principium animalis secundum quod animal, sed considerat sperma, et hoc sufficit.

Ad rationes.

Ad 1.1 Ad primam, dico quod bene concedo quod definitio substantie est sufficiens cognitio eius in genere cognitionis quidditative, et illa non requirit omnes causas, et ideo completa est solum in tali genere, sed non est perfecta simpliciter, cum non consideret omnes causas. Et hinc est quod nulla scientia specialis facit scire rem perfecte, et ideo dependent a scientia una universali, cuiusmodi est metaphysica.

Ad 1.2 Ad aliam rationem, dico quod medicina non considerat sanitatem secundum quod ens, sed secundum quod sanitas, et ideo non oportet quod cognoscat omnes causas eius; nec oportet quod cognoscat primam ideam sanitatis, cum non consideret eam in quantum est ens.

Ad 1.3 Ad tertiam, dicendum quod demonstrationes mathematice faciunt scire per omnes causas illius; ille enim res de quibus sunt scientie mathematice non habent omnes causas, non enim habent causam materialem

---

[a] Aristoteles, *Met.*, III, 1, 995 a 24–b 4

---

18 si] *s. l.* M

3 To this question I reply that it is necessary for the cognition of a thing that has every cause to cognize every cause, both extrinsic and intrinsic, since knowing is cognizing through the cause. So insofar as one lacks cognition of the cause, to that extent one lacks knowledge of the thing.

Furthermore, this is obvious from Aristotle in *Metaphysics* III, who maintains that a complete cognition is a resolution of every doubt, but if we knew one cause then it would be possible to doubt the others. Therefore there is no complete cognition unless every cause is cognized.

Furthermore, this is obvious since we do not know a thing until we know how to resolve it into all its principles, but we know this only if we know every cause; therefore etc.

But it should be understood that since any knowledge considers a thing under some account at the level of its species, it suffices that it consider the causes of that thing according to its own kind. For instance, a doctor considers health with respect to health, but not with respect to being. Under that account, then, it is not necessary for the doctor to go all the way to a cognition of the first cause, since this is not a cause of health considered as health, but a cause of health considered as being. In the same way, the Philosopher in *De animalibus* ought not to consider prime matter, since this is not a principle of an animal as animal. Instead, he considers sperm, and this suffices.

In response to the arguments.

Ad 1.1 To the first I reply that I certainly concede that the definition of a substance is a sufficient cognition of it in the genus of quidditative cognition, and that does not require every cause, and so it is complete only within such a genus. But it is not perfect without qualification, since it does not consider every cause, and for this reason no special science produces perfect knowledge of a thing. And so ⟨all the special sciences⟩ depend on one universal science, which is metaphysics.

Ad 1.2 To the second I reply that medicine does not consider health according to being, but as it is health, and so it is not necessary that it cognize every cause of health. Nor is it necessary that it cognize the first idea of health, since it does not consider it as it is a being.

Ad 1.3 To the third it is to be replied that mathematical demonstrations produce knowledge through every cause.[6] For those things of which there is mathematical knowledge do not have every cause, for they do

---

[6] That is, mathematical demonstrations produce knowledge through every cause relative to the thing demonstrated, not every cause absolutely, since mathematical entities lack all but a formal cause.

nec efficientem nec finalem, ut dicit Aristoteles[a] III *Metaphysice*, quod in mathematicis non est finis nec bonum; solum ergo considerat causam formalem, unde non facit scire rem nisi per causam formalem. Unde ⟨et⟩si non faciant scire rem nisi per unam causam, tamen sunt certissime, quia non requiritur aliqua alia causa ad demonstrationem illam.

⟨Questio 12⟩

Consequenter queritur utrum demonstratio sit ex veris.

1 Et arguitur quod non, quia demonstratio per impossibile est demonstratio; et non procedit ex veris, quia procedit ex hypothesi falsa concludendo falsum; ergo etc.

2 Oppositum vult Philosophus.[b]

3 Ad questionem istam, dico quod demonstratio est ex veris, quia quod non est non contingit scire; sed falsum non est; ergo etc. Tunc arguo: conclusio que scitur est vera; et verum non scitur nisi ex veris; ergo etc. Sed quia verum, secundum Lincolniensem,[c] dividitur in contingens et necessarium, oportet quod demonstratio sit ex veris necessariis, quia omnis demonstratio est syllogismus faciens scire; sed solum necessarium scitur et non nisi ex necessariis; ergo etc. Sed cum demonstratio sit ex necessariis, non solum oportet quod sit ex necessariis simpliciter, sed quoad nos, quia demonstratio est syllogismus scientificus per quem scimus; sed non scimus nisi ex nobis notis; ideo oportet quod demonstratio sit ex necessariis simpliciter et nobis. Talis demonstratio est mathematica, et ideo talis demonstratio est potissima. Sed si aliqua procedat ex nobis notioribus solum, non est demonstratio simpliciter, nec demonstratio procedens ex notioribus nature solum, sed solum illa que procedit ex notioribus nobis et nature; talis solum est in mathematicis.

---

[a] Aristoteles, *Met.*, III, 2, 996 a 21–34   [b] Aristoteles, *AnPo*, I, 2, 71 b 20–22   [c] Robertus Lincolniensis, *In Posteriorum analyticorum libros* I, 4 (ed. Rossi, p. 110)

---

16 mathematica]   mai[ca] M

not have a material, efficient, or final cause. For instance, Aristotle says in *Metaphysics* III that in mathematics there is no end nor any good. Therefore it considers the formal cause alone. Hence it does not produce knowledge of a thing except through the formal cause. Hence even if mathematical demonstrations produce knowledge of a thing through just one cause, still they are the most certain demonstrations, since no other cause is required for this sort of demonstration.

Question 12.

Next it is asked whether demonstration arises from true premises.

1 It is argued that it does not, since *per impossibile* demonstration is demonstration, and it does not proceed from true premises, since it proceeds from a false hypothesis in order to conclude that something is false. Therefore etc.

2 The opposite view is held by the Philosopher.

3 To this question I reply that demonstration is from true premises, since one cannot know what is not, but what is false is not, therefore etc. Then I argue that the conclusion that is known is true, and what is true is not known except through truths, therefore etc. But since the true, according to Grosseteste, is divided into the contingent and the necessary, demonstration must be from necessary truths. For every demonstration is a syllogism that produces knowing, but only the necessary is known, and it is not known except from what is necessary; therefore etc. But since demonstration is from necessary premises, it must be not only from necessary premises without qualification, but from premises necessary in relation to us. For a demonstration is a knowledge-producing syllogism through which we know, but we do not know except from what is known to us. Therefore demonstration must be from necessary truths without qualification, and in relation to us. Mathematical demonstrations are like this, and therefore they are the highest sort of demonstration. But if anyone proceeds from what is better known to us alone, this is not a demonstration without qualification, nor does a demonstration proceeding from what is better known in its nature alone count as a demonstration without qualification. The only one that counts is one that proceeds from what is better known to us and in its nature. Such demonstrations occur only in mathematics.

Ad 1 Ad rationem, dico quod in demonstratione ad impossibile est triplex processus: primus est ex falsa hypothesi cum quodam vero coassumpto concludere falsum, et iste est syllogisticus; secundus | est ex interemptione illius falsi ad interemptionem hypothesis; tertius est ex interemptione hypothesis ad veritatem propositi, et iste ultimus processus est demonstrativus et procedit secundum illam maximam 'non de eodem simul esse et non esse', et secundum istum processum procedit ex veris.

M 101rb

⟨Questio 13⟩

Consequenter queritur circa partem illam: *priora autem ad posteriora* (71 b 34), ubi comparat singulare singularia ad universalia, et ibi queritur utrum universalia sint nobis notiora aut singularia.

1.1 Et arguitur quod universalia, quia illud ex quo procedimus naturaliter ad cognitionem aliorum est nobis notius; sed universalia sunt huiusmodi, ut patet per Aristotelem[a] I *Physicorum*.

1.2 Item, homo est illud quod est per intellectum, ergo illud quod est nobis notius per intellectum est nobis notius simpliciter; sed universalia sunt huiusmodi; ergo etc.

2 Oppositum arguitur: illud quod est propinquius sensui est nobis magis notum; sed singularia sunt huiusmodi; ergo etc.

3 Ad istam questionem dico quod singulare dicitur dupliciter: singulare simpliciter et singulare in respectu; singulare simpliciter quod non predicatur de aliis sed alia de ipso, singulare in respectu quod predicatur de aliis et alia de ipso, sicut sunt species et genera intermedia. Unde ad questionem dico quod singularia simpliciter sunt nobis magis nota quam universalia, quia cognoscuntur cognitione sensitiva; universalia autem non sic cognoscuntur; ideo sunt nobis minus nota.

Sed est intelligendum quod licet singularia absolute loquendo sint nobis notiora quam universalia, tamen universalia sunt nobis magis nota secundum intellectum, quia illud quod per se cognoscitur ab aliqua virtute

---

[a] Aristoteles, *Phys.*, I, 1, 184 a 16–26

---

21 triplex] falsus M   16 singularia] *i. m. et post* nota M

Ad 1 In response to the argument, I reply that in a demonstration *ad impossibile* the process has three stages. The first stage concludes from a false hypothesis, together with a certain true assumption, that something is false, and this is syllogistic. The second moves from the destruction of that falsehood to the destruction of the hypothesis. The third proceeds from the destruction of the hypothesis to the truth of what was proposed, and this last stage is demonstrative, and proceeds according to the general principle that the same thing cannot both be and not be simultaneously. And demonstration at this stage proceeds from true premises.

Question 13.

Next we ask about the phrase *'prior to posterior'*, where he compares singulars to universals, and here we ask whether it is universals or singulars that are better known to us.

1.1 It is argued that universals are better known to us, since that from which we proceed naturally to cognition of other things is more known to us, but universals are like this, as is apparent from Aristotle, *Physics* I.

1.2 Furthermore, human beings are what they are through their intellects. Therefore what is better known to us through intellect is better known without qualification. But universals are of this sort; therefore etc.

2 On the other hand, it is argued that what is nearer to the senses is better known to us, but singulars are of this sort; therefore etc.

3 To this question I reply that 'singular' is said in two ways, singular without qualification and singular in a certain respect. The singular without qualification is what is not predicated of others, but others of it. The singular in a certain respect is what is predicated of others and others of it, as intermediate species and genera are. So I reply to the question that singulars without qualification are better known to us than universals are, since they are cognized by sensory cognition, whereas universals are not cognized in that way, and for this reason they are not as well known to us.

But it should be understood that although singulars are, speaking absolutely, better known to us than universals, still universals are better known to us with respect to the intellect, since what is cognized *per se* by a power that is in us is better known with respect to that power than what is cognized accidentally by it. But universals are cognized *per se* by

que est in nobis notius est secundum illam virtutem quam illud quod per accidens; sed universalia per se cognoscuntur ab intellectu, singularia per accidens, quia per reflexionem; ergo etc. Et sicut in intellectu quanto aliqua sunt magis universalia tanto citius percipiuntur ab intellectu, sic[ut] singularia magis confusa citius percipiuntur a sensu. Unde Avicenna[a] dicit quod prius percipitur de homine a longe quod sit hoc ens quam quod sit hoc animal, et quod sit hoc animal quam hic homo. Unde dicit Philosophus[b] in principio *Physicorum* quod in principio vocat puer omnes viros patres et omnes feminas matres, et postea discernit patrem proprium ab alio. Et causa huius est quia intellectus noster vadit de potentia ad actum, et similiter sensus, et ideo illud quod magis confusum est prius comprehenditur a sensu et ab intellectu, cum sit magis in potentia.

Sed tu queres: cum singulare sit magis cognitum sensui et universale intellectui, quod illorum est prius? Dico quod singulare prius est secundum esse reale quam universale sit in intellectu, quia universale aut nihil est aut posterius est, cum omnis communitas a singularitate procedat.

Ad 1.1 Ad rationem, dico quod a cognitione universalium procedimus in cognitionem singularium in respectu et non in cognitionem singularium simpliciter, sed ex cognitione singularium simpliciter procedimus in cognitionem universalium. Unde hoc quod dicit Philosophus I *Physicorum* intelligendum est comparando universale ad singulare ⟨in⟩ respectu, non autem est verum comparando universale ad singulare simpliciter.

Ad 1.2 Ad aliam rationem, dico quod verum concluderet, si cognitio intellectiva in nobis non esse⟨t⟩ ex alia cognitione, sed nunc ex alia cognitione est.

⟨Questio 14⟩

Consequenter queritur utrum magis oporteat credere conclusioni quam principiis.

1 Et videtur quod conclusioni magis, quia illi magis credendum est cuius cognitio certior est; sed cognitio conclusionis est certior, cum sit per causam, et alia per experimentum, quod fallax est; ergo etc.

---

[a] Avicenna, *Sufficientia* I, 1 (1508, ff. 13rb–va)   [b] Aristoteles, *Phys.*, I, 1, 184 b 11–13

---

39 singularium[1]] universalium M

the intellect, and singulars accidentally, since they are cognized through reflection; therefore etc. And just as in the intellect the more universal is more quickly perceived by the intellect, so more confused singulars are more quickly perceived by the senses. Whence Avicenna says that it is perceived of a human being at a distance that he is *this being* before it is perceived that he is *this animal*, and it is perceived that he is *this animal* before it is perceived that he is *this human being*. So the Philosopher says in the beginning of the *Physics* that a child at first calls every man father and every woman mother, and afterwards distinguishes his own father from another. And the reason is that our intellect goes from potentiality to actuality, and the senses likewise, and so both sense and intellect first grasp that which is more confused, since it is in greater potentiality.

But you will ask: since the singular is more cognized by the senses and the universal by the intellect, which of these is prior? I reply that in the intellect the singular is prior to the universal in real being, since the universal is either nothing or is posterior, for every commonality proceeds from singularity.

Ad 1.1 In response to the argument, I reply that from a cognition of universals we proceed to the cognition of singulars in a certain respect, and not to the cognition of singulars without qualification. But from a cognition of singulars without qualification we proceed to the cognition of universals. Accordingly, what the Philosopher says in *Physics* I is to be understood as comparing a universal to a singular in a certain respect. But it is not true when comparing a universal to a singular without qualification.

Ad 1.2 To the second argument, I reply that it would conclude truly if an intellectual cognition in us were not from another cognition, but in fact it is from another cognition.

Question 14.

Next it is asked whether one ought to believe the conclusion more than the principles.

1 It seems that one ought to believe the conclusion more, since that is more to be believed whose cognition is more certain, but the cognition of the conclusion is more certain since it occurs through the cause, whereas the other occurs through experience, which is fallible; therefore etc.

2 Oppositum arguitur sic: certior est illa cognitio, et notior, que non est per collationem quam illa que est per collationem, quia per illam assimilatur substantiis separatis; sed cognitio principiorum non est per collationem, cognitio conclusionis per collationem; ergo etc.

3 Ad istam questionem dico quod loquendum est de conclusione dupliciter: aut postquam probatur aut ante. Si ante, non magis credendum est conclusioni quam principiis, cum ipsa sit ignota et principia sint prescita, quia nec sciens nec melius dispositus quam sciens magis credit eis que non novit quam eis que novit.

Et tu dices: proba mihi quod magis credendum sit principiis quam conclusioni postquam conclusio demonstratur. Probo istud sic: verius habet illud esse quod habet esse ex se quam illud quod habet esse ex alio, quia eadem est dispositio in esse et veritate, ergo verius sciuntur illa que sciuntur ex se ipsis quam ea que ex aliis; sed cognitio principiorum est ex se ipsis, cognitio conclusionum ex aliis; ergo etc.

Ad 1 Ad rationem in oppositum, concedo maiorem et nego minorem, quia certitudo de cognitione conclusionis accipitur ex principiis, ergo principia sunt magis certa.

Et tu dicis quod cognitio conclusionis est per causam. Dico quod ex hoc probo quod cognitio principiorum est certior, quoniam causa cognitionis conclusionis est ex principiis, et qui errat in principiis errat in conclusione et non e converso; ideo etc.

Sed tu dices: experimentum, quod est via cognoscendi principia, est fallax. Dico quod duplex est experimentum: unum ex quo accipitur cognitio universalium in speculabilibus, ⟨aliud in agibilibus⟩; unde experimentum in agibilibus est fallax, in speculabilibus autem non, et ex tali experimento habetur cognitio principiorum.

---

17 quia] sed M

2 On the other hand, it is argued thus: a cognition that does not occur through a collecting together of premises is more certain and more known than is one that occurs through a collecting together of premises, since the former happens in the way it does in the separated substances. But the cognition of principles is not through a collecting together of premises, and the cognition of a conclusion is through a collecting together of premises; therefore etc.

3 I reply to this question that a conclusion should be spoken of in two ways, either after it is proven or beforehand. If beforehand, the conclusion is not more to be believed than the principles, since it is unknown then, and the principles are known beforehand. For neither someone with knowledge nor someone in a state superior to knowledge believes the things he does not know more than he believes the things he knows.

You will reply: 'prove to me that principles are more to be believed than the conclusion *after* the conclusion is demonstrated'. I prove it thus: what has being from itself has being more truly than that which has being from another, because a thing's state of being is the same as its state of truth. Therefore those things that are known from themselves are more truly known than those known from others. But the cognition of principles is from themselves and the cognition of conclusions is from others. Therefore etc.

Ad 1 In response to the argument in opposition, I grant the major and deny the minor, for certitude regarding cognition of the conclusion is received from principles, so principles are more certain.

You say that cognition of the conclusion occurs through the cause. I reply that I can prove from this that the cognition of principles is more certain, since the cause of the cognition of the conclusion is from principles, and so whoever goes astray in the principles goes astray in the conclusion, but not conversely, therefore etc.

But you will say experience, which is the path to the cognition of principles, is fallible. I reply that experience is two-fold. One is the experience from which a cognition of universals is received in speculative thought, and the other is that from which a cognition of universals is received in action. So, experience in action is fallible, but in speculative thought it is not. And we have a cognition of principles from the latter sort of experience.

⟨Questio 15⟩

Consequenter queritur utrum omnium contingat esse scientiam per demonstrationem.

1.1 Et arguitur quod sic, quia omne quod est vel est causa vel est effectus; sed effectus demonstratur per causam et e converso; ergo etc. Maior | patet, quia causa et effectus ex opposito dividunt ens. Minor patet, quia effectus demonstratur per causam propter quid et causa per effectum quia.

M 101va

1.2 Item, si contingat circulariter demonstrare, omnium potest esse scientia per demonstrationem, quia sic posteriora possunt demonstrari per priora et e converso. Nunc autem contingit circulariter demonstrare, per Aristotelem[a] secundo huius, ut: si est pluvia, necesse est terram esse madidam, et si terra est madida, necesse est fieri vaporem, et si vapor fit, necesse est fieri pluviam, ergo a primo, si pluvia est, pluvia est.

2 Oppositum arguitur sic, quia si omnia sciuntur per demonstrationem, nihil sciretur; sed hoc est falsum; ergo et primum. Probatio antecedentis, quia si omnia scirentur per demonstrationem, tunc non essent aliqua simpliciter prima, quia omnia essent demonstrabilia, et sic non contingeret scire prima, nec per consequens posteriora.

3 Dicendum ad hoc quod non omnium est scientia per demonstrationem, quia nec omnium est scientia simpliciter que habetur per demonstrationem, quia primorum principiorum non est scientia simpliciter. Non enim omnis cognitio est scientia, ut cognitio sensitiva, quia tunc bruta scirent; item, nec omnis cognitio intellectiva, ut incomplexi, immo magis intellectus; item, non omnis cognitio complexi, sed que est per causam. Et ideo quorum non est causa non est scientia simpliciter; sed primorum principiorum non est causa; ergo eorum non est scientia simpliciter. Sed tamen cognitio eorum est principium scientie, quia ex cognitione principiorum devenimus in cognitionem conclusionis. Quare non omnium est scientia per demonstrationem. Licet tamen principiorum non sit scientia simpliciter, tamen eorum est scientia per accidens, quia scientia per accidens est

---

[a] Aristoteles, *AnPo*, II, 12, 95 b 38–96 a 8

---

11 fit] sit M

Question 15.

Next it is asked whether there can be knowledge through demonstration of everything.

1.1 It is argued that there can be, for everything that is is either a cause or an effect. But an effect is demonstrated through the cause, and conversely; therefore etc. The major premise is evident since cause and effect divide being into opposites. The minor premise is evident since the effect is demonstrated through the cause that gives the reason why, whereas the cause is demonstrated through the effect that shows that it is so.

1.2 Furthermore, if circular demonstration is possible, then there can be knowledge through demonstration of everything that is known, for thus the posterior can be demonstrated through what is prior, and conversely. But it is possible to demonstrate in a circle according to Aristotle in the second book of this treatise. For instance, if there is rain, it is necessary for the earth to be soaked, and if the earth is soaked, it is necessary that vapor arises, and if vapor arises, then it is necessary that there be rain. Therefore, from first ⟨to last⟩, if there is rain then there is rain.

2 The opposing view is argued for as follows: If everything is known through demonstration then nothing will be known. But this second statement is false, therefore the first is also. Proof of the antecedent: If everything that is known is known through demonstration then there is nothing that is first without qualification, since everything would be demonstrable, and thus it is not possible to know first things, nor, consequently, what is posterior.

3 One should reply to this that there is not knowledge through demonstration of everything, since there is not knowledge without qualification of everything that is gotten through demonstration, since there is no knowledge without qualification of first principles. The reason for this is that not every cognition is knowledge. For instance, sensory cognition is not, for then the beasts would know. Furthermore, neither is every intellectual cognition knowledge. For instance, cognition of a single term is instead understanding. Moreover, not every cognition of a proposition is knowledge, but only that which is reached through its cause. And so there is no knowledge without qualification of what has no cause. But there is no cause for first principles. Therefore there is no knowledge without qualification of first principles. But nonetheless cognition of these

30 quando aliquid scitur per accidens; nunc principiorum est scientia per accidens, quia non habent causam; ideo eorum cognitio est per accidens.

Ad 1.1 Ad argumentum. Ad maiorem, dico quod vera est. Ad minorem, dico quod verum est quod effectus demonstratur per causam simpliciter, et ideo scitur simpliciter, sed causa demonstratur per effectum
35 demonstratione quia, et ideo scitur secundum quid.

Ad 1.2 Ad aliud, Philosophus[a] negat maiorem, quia non omnium est scientia per demonstrationem; non enim contingit demonstrare circulariter omnia, sed solum convertibilia. Potest tamen dici ad minorem quod non contingit circulariter demonstrare simpliciter.

40 Ad Philosophum secundo huius Albertus[b] docet respondere et dicit quod in illis que generantur circulariter non potest fieri circulatio ita quod [non] sit reditio ab eodem in idem secundum numerum, sed secundum speciem, et propter hoc non est in eis circulatio simpliciter. Nec pluvia hesterna est prior secundum naturam pluvia hodierna, cum circulatio
45 illa fiat in infinitum et in infinito non est prius, quia ibi non est primum, et quodlibet prius dicitur respectu alicuius primi, et ideo non est pluvia prior altera secundum naturam.

⟨Questio 16⟩

Consequenter queritur utrum hec conditio 'per se' sit possibilis in entibus.

1.1 Et arguitur quod non, quia 'per' denotat causam et 'se' est pronomen reflexivum, denotat ergo aliquid esse causam sui ipsius; sed hoc est
5 impossibile in entibus; ideo etc.

---

[a] Aristoteles, *AnPo*, II, 12, 95 b 38–96 a 2   [b] Albertus Magnus, *In Analytica posteriora*, II, III, 7 (ed. Borgnet, p. 207)

---

41 ita]   natura M   4 reflexivum]   reciprocum M

is the starting point of knowledge, since we arrive at cognition of the conclusion from a cognition of the principles. So there is not knowledge of everything through demonstration. Yet even if there is no knowledge without qualification of principles, still there is accidental knowledge of principles, since accidental knowledge occurs when something is known accidentally. And the knowledge of principles is accidental, since they have no cause, therefore cognition of them is accidental.

Ad 1.1 In response to the argument, as regards the major premise, I reply that it is true. As regards the minor premise, I reply that it is true that the effect is demonstrated through the cause without qualification and is therefore known without qualification. But the cause is demonstrated through the effect by demonstration that it is so, and therefore is known only in a certain respect.

Ad 1.2 In response to the other, the Philosopher denies the major premise, since there is not knowledge of everything through demonstration. For it is not possible to demonstrate everything in a circle, but only what is convertible. Still, one can reply to the minor premise that it is not possible to demonstrate in a circle without qualification.

To the Philosopher in the second book of this treatise, Albert the Great teaches how to respond. He says that, among things that are generated in a circle, a circle cannot arise such as to yield a return from what is numerically the same into itself. Instead, the circle concerns what is the same in species. And because of this, there is no circularity without qualification. Nor is yesterday's rain prior by nature to today's rain, since the circularity keeps going into infinity, and in an infinite there is no prior, since there is no first, whereas whatever is said to be prior is said to be so with respect to some first. And so there is no rain prior to another rain by nature.

Question 16.

Next it is asked whether the state of being *per se* is possible for existing things.

1.1 It is argued that it is not, since '*per*' indicates cause, and '*se*' is a reflexive pronoun. '*Per se*', then, indicates that something is the cause of itself. But this is impossible for existing things. Therefore etc.

1.2 Item, per se dicitur esse illud quod non habet causam aliam, quia privat causalitatem respectu alterius; sed nihil est quod non habet causam aliam nisi primum; ergo etc.

2 Oppositum arguitur sic: iste tres conditiones se habent secundum ordinem: 'de omni', 'per se', et 'primo'; sed ista conditio 'primo' est possibilis in entibus; ergo et 'per se'.

3 Ad istam questionem, dico quod 'per se' potest dupliciter intelligi: vel privative vel positive. Si privative, contingit esse quattuor modis: uno modo per privationem cuiuslibet causalitatis, et hoc modo sola causa prima est per se; solum enim primum est illud quod non est effective ab alio et omnia sunt effectus eius, unde secundum quod dicit privationem cause efficientis sola causa prima est per se. Secundo modo, secundum quod 'per se' dicit privationem cause materialis, et sic separata omnia sunt per se. Tertio modo, [materialis] dicitur aliquid esse per se, quia non est per aliud sicut per subiectum, et sic omnis substantia dicitur esse per se, quia nulla substantia indiget alio tamquam subiecto. Istos tres modos ponit Lincolniensis.ᵃ Quartum modum possumus addere, quod aliquid dicitur per se quia secundum se est tale, ita quod ipsum non est tale per causam aliam ⟨formalem⟩ nec per causam aliam effectivam, loquendo de causa proxima efficiente, et hoc modo dicuntur omnes propositiones per se in primo modo et quarto; in illis enim subiectum participat predicatum non per aliud formaliter nec effective proximum, ut in primo modo homo est animal non per aliam causam formalem, similiter in quarto interfectum interiit non per aliam causam effectivam proximam.

Alio modo potest 'per se' sumi positive, et hoc duobus modis: uno modo, quia significat illud cui additur esse causam sui ipsius, et hec conditio ⟨non⟩ est possibilis in entibus, quia nihil est causa sui ipsius, cum causa sit ad cuius esse sequitur aliud, et idem non est aliud a se ipso, etiam causa est prius causato, et idem non est prius se ipso; alio modo dicitur aliquid per se quia aliquid habet in se causam quare aliud ei inest, et ista conditio est 'per se' | in secundo modo.

M 101vb

ᵃ Robertus Lincolniensis, *In Posteriorum analyticorum libros*, I, 4 (ed. Rossi, p. 111)

---

6 esse] *i. m.* M ‖ habet] privat *ante corr.* M   17 cause efficientis] *inv.* M   21 subiecto] subiectum M   26 quarto] 4 M   27 proximum] proximo M   28 similiter] simpliciter M   33 idem] illud M   36 est] de M

1.2 Furthermore, that is said to be *per se* which does not have another for a cause, since it lacks causality with respect to another, but there is nothing that does not have another for a cause except the first cause, therefore etc.

2 On the other hand, it is argued thus: these three states stand to one another in this order: universal (*de omni*), *per se* and primary. But the state of being primary is possible in things that exist, and so *per se* is too.[7]

3 As regards this question, I maintain that *per se* can be understood in two ways, privatively or positively. If privatively, there are four possible ways: in one way through privation of every causality, and in this way only the first cause is *per se*, for only the first is something that is not an effect of another, and everything else is an effect of the first. So according to that sense that indicates privation of an efficient cause, only the first cause is *per se*. In the second way *per se* indicates privation of material cause, and in this way every separated substance is *per se*. In the third way something is said to be *per se* that is not because of another in the way that something is because of a subject, and in this way every substance is said to be *per se*, since no substance needs another as a subject. Grosseteste gives these three ways. We can add a fourth way in which something is called *per se*: because it is of such a sort in itself, so that the thing itself is not of such a sort because of another formal cause, nor because of another efficient cause, speaking of the proximate efficient cause.[8] This is the way in which all propositions are called *per se* in the first and fourth modes. For in these modes the predicate belongs to the subject neither because of another formally nor because of an efficient proximate cause. For instance, in the first mode a human being is an animal not because of another formal cause, and similarly in the fourth mode someone who is killed perishes not because of another efficient proximate cause.

In another way '*per se*' can be taken positively, and here there are two ways. One is because it signifies that to which being its own cause is added. And this condition is not possible in existing things since nothing is a cause of itself, for a cause is that upon which the being of *another* follows, and the same thing is not other than itself. Also, a cause is prior to the thing caused, and the same thing is not prior to itself. In the second way a thing is called *per se* because it has in itself the cause for why another inheres in it, and this state is the second mode of being *per se*.

---

[7] That is, each is a narrower category contained in the categories before it in the list, so that if any one is possible, then those listed before it are, too.

[8] Simon refers here to the fourth way of being *per se* introduced in chapter four of the *Posterior Analytics* (73a34ff). According to Aristotle, something is said to be *per se* when: (1) it is included as a predicate in the definition of the subject; (2) it is a subject included as an element in the definition of its predicate; (3) it is predicated of nothing else as a subject; or (4) it is a necessary consequence of a subject's essence.

Ad rationes patet; procedunt enim viis suis.

Ad 1.1 Ad primam, dico quod ⟨'per se'⟩ vel denotat aliquid esse principium sui ipsius vel aliquid habere in se principium mediante quo aliud ei inest, et isto ultimo modo est hec conditio 'per se' possibilis.

Ad 1.2 Ad aliam, dico quod omne quod est preter primum habet causam remotam, ad minus, quedam tamen sunt que non habent causam proximam.

⟨Questio 17⟩

Consequenter queritur utrum ea que pertinent ad quod quid est insint per se primo modo.

1.1 Et arguitur quod non, quia materia et forma pertinent ad quod quid est substantie sensibilis; sed materia non inest forme per se primo modo, nec forma materie inest per se primo modo.

1.2 Item, genus et differentia pertinent ad quod quid est speciei; nunc autem genus de differentia non predicatur primo modo; ergo etc.

2 Oppositum arguitur: que pertinent ad quod quid est aut insunt primo modo aut secundo; non secundo, quia tunc propria passio pertineret ad quod quid est sui subiecti; ergo que pertinent ad quod quid est insunt per se primo modo.

3 Ad istam questionem, dico quod illa que pertinent ad quod quid est possunt referri ad se invicem vel ad illud cuius sunt quod quid est. ⟨Si referantur ad illud cuius sunt quod quid est⟩, sunt in natura et substantia rei, ergo etc. Sed si referantur inter se, non insunt per se primo modo, quia illa que insunt primo modo, unum ponitur in definitione alterius, sed illa que pertinent ad quod quid est, unum non ponitur in definitione alterius, quia sic omnis definitio esset nugatoria. Probatio, quia quando aliqua duo adiunguntur ad invicem immediate et unum est in ratione alterius est nugatio; sed hoc contingeret si unum pertinentium ad quod quid est poneretur in definitione alterius; ergo etc. Tamen oportet quod, cum aliqua duo referantur ad quod quid est alicuius, que sunt de ipso quod quid est, quod unum sit in potentia et reliquum sit actus. Unde dicit logicus quod genus est ut materia, differentia vero ut forma.

The answers to the arguments are obvious, for they each hold in their own way.

Ad 1.1 To the first I reply that *'per se'* indicates either something's being its own principle or something's having in itself the principle by which another inheres in it. And in this last way this state of being *per se* is possible.

Ad 1.2 To the other I reply that everything aside from the first has a remote cause, at least, but there are some things that do not have a proximate cause.

Question 17.

Next it is asked whether those things that pertain to a thing's essence (*quod quid est*) inhere *per se* in the first mode.

1.1 It is argued that they do not, for matter and form pertain to the essence of a sensible substance, but matter does not inhere in the form *per se* in the first mode, nor does the form inhere in matter *per se* in the first mode.

1.2 Furthermore, genus and difference pertain to the essence of the species. Now the genus is not predicated of the difference in the first mode. Therefore etc.

2 The opposing view is argued thus: what pertains to the essence inheres in either the first mode or the second. Not in the second, since then a proper attribute would pertain to the essence of its subject. Therefore what pertains to the essence inheres *per se* in the first mode.

3 To this question I reply that the things that pertain to a thing's essence can be referred to one another or to that of which they are the essence. If they are referred to that of which they are the essence, then they are in the nature and substance of the thing. Therefore etc. If, on the other hand, they are referred to one another then they do not inhere *per se* in the first mode, since of those which inhere *per se* in the first mode one is contained in the definition of the other. But of those which pertain to the essence, one is not contained in the definition of the other, since if it were then every definition would be nugatory. Proof: for when two things are joined to one another immediately and one is in the account (*ratio*) of the other, it is nugatory. But this would be the case if one of the things pertaining to the essence were contained in the definition of the other, therefore etc. But it is necessary, when two things are referred to a thing's

Ad rationes patet, quia procedunt in equivoco.

⟨Questio 18⟩

Consequenter queritur utrum genus de differentia predicetur primo modo dicendi per se. Secundo queratur utrum hec sit per se 'homo est homo'. Tertio utrum 'animal est homo'. Quarto utrum hec 'Callias est homo'.

1.1 Et primo arguitur quod sic: genus predicatur de differentia, aut ergo predicatur per se aut per accidens. Si per se, habetur propositum. Si per accidens, aut quia subiectum accidit predicato aut predicatum subiecto, quod non contingit, quia sunt idem secundum rem, aut quia utrumque accidit alicui tertio, quod non est verum, quia utrumque significat substantiam et substantia nulli accidit.

1.2 Item, quecumque aliqua duo sunt eadem omnino, quidquid predicatur de uno, et de alio et eodem modo; sed genus, species et differentia sunt huiusmodi, quia per Aristotelem[a] in VII *Metaphysice* differentia ultima est totum esse speciei; sed genus predicatur per se de specie; ergo etc.

1.3 Item, nullum animal est rationale per se; omnis homo est rationalis per se; ergo etc.; conclusio est falsa, et non minor, ergo maior; ergo sua contradictoria est vera per se, scilicet hec 'rationale est animal'.

2 Oppositum arguitur sic: si concretum de concreto per se, et abstractum de abstracto; si ergo hec sit vera 'rationale est animal' per se, hec erit vera per se 'rationalitas est animalitas'; sed hec est falsa, quia eorum sunt diverse rationes.

---

[a] Aristoteles, *Met.*, VII, 12, 1038 a 25–26

---

5 predicatur] per^{ur} M

essence and are of this essence, that one be in potentiality and the other in act. Thus the logicians say that the genus is like matter whereas the difference is like form.

The response to the arguments is clear since they go through by equivocation.

Question 18.

Next it is asked whether genus is predicated of difference in the first mode of speaking *per se*. Second ⟨q. 19⟩ it is asked whether this is *per se*, 'The human being is a human being'. Third ⟨q. 20⟩ it is asked whether 'The animal is a human being' is per se. Fourth ⟨q. 21⟩ it is asked whether 'Callias is a human being' is *per se*.

1.1 And first it is argued that the genus is predicated of the difference. So then it is predicated either *per se* or *per accidens*. If *per se*, then we have what is proposed. If *per accidens*, then either because the subject is accidental to the predicate or the predicate to the subject, which cannot occur because they are the same thing, or else because both belong to some one third thing, which is not true since both signify substance, and a substance is accidental to nothing.

1.2 Furthermore, whenever two things are wholly the same, whatever is predicated of one is also predicated of the other, and in the same way. But genus, species and difference are of this sort, since Aristotle in *Metaphysics* VII says the last difference is the whole of the species. But the genus is predicated *per se* of the species. Therefore etc.

1.3 Furthermore, no animal is rational *per se*, every human being is rational *per se*, therefore etc. The conclusion is false and the minor premise is not, so the major premise is false. Therefore its contradictory is true *per se*, namely 'A rational thing is an animal'.

2 The opposite view is argued thus: if the concrete is predicated *per se* of the concrete, the abstract is also predicated *per se* of the abstract. If, then, this is true, 'A rational thing is an animal', *per se*, this will be true *per se*, 'Rationality is animality'. But this is false since there are different accounts for these.

3 Ad istam questionem intelligendum duo: primum est quod genus et differentia non sunt diversa secundum naturam et rem; secundum est quod genus et differentia diversificantur secundum rationem.

Probatio primi, quia substantia rei simplex est, ergo illa que pertinent ad essentiam rei debent importare simplicem essentiam. Dico autem 'simplicem', non quia simplex est omnino, sed quia non est partita.

Item, dicit Philosophus[a] quod in substantia rei non est ordo, puta non differre in homine unam formam per quam est substantia et alia⟨m⟩ per quam est animal et homo, quia si sic, forma [differre] esset accidentalis. Et ratio huius est, quia si forma substantialis adveniat materie, quantumcumque fuerit ⟨in⟩completa, dat esse in actu materie, quia dat esse tale per quod materia potest manere in actu per speciem, sicut patet de forma elementi, que est maxime incompleta. Nunc autem omnis forma adveniens enti in actu secundum speciem est accidentalis. Per Commentatorem[b] III *De anima*, differentia est inter formam substantialem et accidentalem; forma enim substantialis est que dat esse simpliciter et primo, forma accidentalis accipit esse simpliciter ab eo cui advenit, et dat esse tale. Modo si forma aliqua adveniat enti in actu secundum speciem, non dat esse primo et absolute ipsi cui advenit, ergo dat sibi esse tale, ergo sequetur quod forma substantialis posterior est accidentali⟨s⟩.

Sed tu dices: licet non adveniat enti in actu completo, potest tamen advenire enti in actu incompleto. Quero a te: aut illud adveniens dat esse primo et absolute aut non. Sed constat quod non, quia primo obtinuit formam incompletam; ergo dat esse solum completum, et illud est esse tale. Et quod dicunt, formam completam et incompletam, verum est quod comparando formas inter se una est completior et alia incompletior, loquendo tamen de forma in se non est ita incompleta | quin det esse secundum speciem ipsi materie.

M 102ra

Unde dicendum quod genus et differentia dicunt unum secundum rem. Et hoc patet, quia, si genus importet unam naturam communem utrique speciei, et differentia essentiam additam, [quia] aut genus dicit totam essentiam utriusque speciei, quod non potest, ⟨aut partem⟩. Si dicas quod

---

[a] Aristoteles, *Met.*, VII, 12, 1038 a 33   [b] Cf. Averroes, *Commentarium magnum in Aristotelis De anima libros*, II, 1, comm. 2 (ed. Crawford, pp. 130–31)

---

25 simplex]   semel M

BOOK I 63

3 Two things should be understood with respect to this question. The first is that genus and difference are not distinct as natures and things. The second is that genus and difference are different in account.

The proof of the first: since the substance of a thing is simple, what pertains to the essence of a thing must convey a simple essence. But I maintain that it is simple not because it is wholly simple, but because it is not divided.

Furthermore, the Philosopher says that in the substance of a thing there is no order. For instance, in a human being, the one form through which it is a substance and the other through which it is an animal and a man do not differ. For if they did, form would be accidental. And the reason for this is that if a substantial form comes to matter, insofar as it is incomplete it gives being in act to matter, for it gives the sort of being through which matter can remain in act through its species, as is obvious in the case of the form of an element, which is the most incomplete sort of form. But every incoming form of a being in act with respect to its species is accidental. As the Commentator says, in *De anima* III, there is a difference between substantial and accidental form, for the substantial form is that which gives being primarily and without qualification. Accidental form receives being without qualification from that to which it comes, and it gives being-such. Now, if some form comes to a being in act with respect to its species, it does not give being primarily and absolutely to that to which it comes. Therefore it gives being-such to it. Therefore it will follow that the posterior substantial form is an accidental form.

But you will reply that although ⟨the difference⟩ does not come to a being that is completed in act, yet it can come to a being in incomplete act. I ask you: Either the incoming form gives being primarily and absolutely or it does not. But it is established that it does not, since it first arrived at something with an incomplete form, therefore it gives only complete being and that is being-such. And when they say that the form is complete and incomplete, it is true that by comparing forms among themselves, one is more complete and another more incomplete. But speaking of the form as it is in itself, it is not in this way incomplete, so that it would not give being with respect to species to its matter.

Whence it must be said that the genus and the difference indicate one thing in reality. And this is clear. For if the genus conveyed one nature common to two species, and the difference an added essence; then either the genus indicates the whole essence of each species, which it cannot do, or else it

genus dicat partem speciei, contra, quia pars de toto non predicatur essentialiter et in quid; genus ergo de specie non predica⟨re⟩tur essentialiter et in quid. Et hec est ratio Avicenne.[a] Tamen genus et differentia differunt secundum rationem, quia immediata adiunctio non differentium secundum rem nec secundum rationem est causa nugationis; si ergo significarent eandem rem sub eadem ratione, quelibet definitio esset nugatoria.

Sed tu dices: quomodo imaginabor ego quod significa⟨n⟩t unam rem secundum diversam rationem? Dico quod in rebus est gradus, sicut in genere substantie est aliqua substantia que non est corporea, alia corporea que non est animata, ita quod in posteriori virtute est prior. Unde in homine est forma qua homo est homo, que virtutem habet in se per quam est substantia et corpus et corpus animatum, et sic de aliis, sicut trigonus in tetragono. Unde ista natura [que] non multiplicatur multiplicatione predicatorum in quid, ⟨sed⟩ secundum diversitatem operationum arguimus diversitatem predicatorum, ut in hoc quod sentit convenit homo cum asino, per hoc quod rationale differt ab aliis. Unde genus significat idem sub ratione indeterminati quod differentia sub ratione determinati. Et sicut est in materia quod ex materia et forma fit unum per essentiam, sic est quod ex genere et differentia fit unum secundum essentiam. Unde ad propositum dico quod genus de differentia non predicatur per se primo modo, quia predicatio per se primo modo exigit quod ratio unius includatur in ratione alterius; nunc autem ratio differentie non includitur in ratione generis nec e converso; ideo etc.

Ad 1.1 Ad rationem primam, dico quod genus predicatur de differentia per accidens, non quia res accidit rei, sed quia ratio accidit rationi, et illud in contrarium procedit si accidet res rei.

Ad 1.2 Ad aliam rationem, dico quod genus, differentia, species sunt idem secundum rem, differunt tamen secundum rationem, quia 'per se' refertur ad rationes subiecti et predicati; quia diversa est ratio speciei et differentie, licet aliquid per se predicetur de specie, non oportet quod per se de differentia predicetur.

[a] Avicenna latinus, *De philosophia prima vel scientia divina*, III, 2 (ed. Van Riet, p. 108)

---

56 et] *al. man.* M   70 indeterminati] -ta *ante corr.* M

indicates a part. If you reply that the genus indicates part of the species, to the contrary, since a part is not predicated of the whole essentially and with respect to the essence, the genus would not be predicated essentially and in some respect of the species. This is Avicenna's argument. But the genus and difference do differ in account, for the immediate joining of things differing neither in reality nor in account is the cause of a nugatory definition. If, then, they signify the same thing under the same account, every definition is nugatory.

But you will reply, 'How am I to imagine that they signify one thing according to different accounts?' I reply that there are degrees in things, for instance, in the genus of a substance. For there is one substance that is not corporeal and another corporeal substance that is not animated, so that the prior is in the posterior virtually. Whence in a human being there is a form by which the human being is a human being, and this form has a power within it through which it is a substance and a body and an animated body, and so on for the rest, just as a triangle is contained in a quadrilateral. Whence the multitude of predicates indicating what the thing is do not multiply its nature; instead, we derive the variety of predicates from the variety of its operations. For instance, inasmuch as he senses, a human being aligns with a donkey, whereas through his being rational he differs from other ⟨animals⟩. Whence the genus signifies under an indeterminate account the same as what the difference signifies under a determinate account. And just as in the case of matter there is one thing in essence that comes from matter and form, so too from genus and difference there comes one thing in essence. So, in response to the question, I say that genus is not predicated *per se* in the first mode of difference, since predication *per se* in the first mode requires that the account of one be included in the account of the other. But the account of the difference is not included in the account of the genus, nor conversely. Therefore etc.

Ad 1.1 To the first argument, I reply that the genus is predicated of the difference accidentally, not because there is one thing that is accidental to another thing, but because the account is accidental to the account, and that the argument for the opposite would result if we had thing accidental to thing.

Ad 1.2 To the second argument I reply that the genus, difference and species are the same thing, but they differ in account. For, *per se* is referred to the accounts of subject and predicate, since the accounts of species and difference are different. Although something is predicated *per se* of a species, it is not necessary that it be predicated *per se* of the difference.

Ad 1.3 Ad aliam rationem, dico quod ibi est fallacia accidentis, quia fallacia accidentis est quando accipiuntur aliqua duo que partim sunt eadem et partim diversa, et accipitur aliquod tertium quod pertinet ad diversitatem eorum, et removetur aliquid ab utroque ipsorum quia removetur ab uno. Et ideo cum rationale sic se habeat respectu animalis et hominis, si removeamus ab homine quia ab animali, est fallacia accidentis.

Item, predictum probatur sic, scilicet quod genus non predicatur per se de differentia: Aristoteles[a] in libro *Topicorum* dicit quod ens non potest esse genus, quia predicatur per se de differentia qualibet primo modo, et genus per ipsum ibi non predicatur de differentia per se. Et hoc est, quia ratio generis includitur in ratione speciei et in ratione differentiarum non, et ideo genus non predicatur per se de differentia nisi mediantibus suis speciebus, et ideo hec est per se: 'rationale, in quantum homo, est animal'.

Item, hoc patet per Philosophum[b] VI *Topicorum*, qui dicit quod, si genus predica⟨re⟩tur per se de differentia, unum animal esset plura animalia. Et hoc patet, quia si rationale per se est animal, rationale per rationem rationalis est animal, et cum ratio rationalis includatur in ratione hominis, homo erit unum animal per rationale. Item, homo est unum animal per hoc quod animal cadit in sua definitione, et animal non cadit in ratione hominis per rationem rationalis, erit tunc homo animal non per rationale, et sic erit aliud animal, et sic unum animal esset duo animalia.

Probatio etiam quod animal non cadit in definitione hominis per rationale, quia tunc esset nugatoria definitio.

Item, si animal caderet in definitione hominis per rationem rationalis, cui non competeret ratio rationalis ei non competeret ratio animalis; cum ergo asino non competat ratio rationalis, ei non competet ratio animalis, quod falsum est.

---

[a] Aristoteles, *Top.*, IV, 6, 127 a 26–34   [b] Aristoteles, *Top.*, VI, 6, 144 a 31–b 2

---

105 aliud] unum *ante corr.* M

Ad 1.3 To the other argument, I reply that it commits the fallacy of accident, for the fallacy of accident occurs when one takes two things that are in part the same and in part different, and then takes a third that belongs to each of them differently, and then removes something from both of them because it is removed from one. So, when 'rational' is related to 'animal' and 'human being' in this way, if we remove 'rational' from 'human being' since we remove it from 'animal', then we have the fallacy of accident.

Further, what has been said, namely that the genus is not predicated of the difference, is proved as follows: Aristotle says in the *Topics* that a being cannot be a genus, since it is predicated *per se* in the first mode of every difference, and the genus, by the argument given there, is not predicated of the difference *per se*. And this is because the account of the genus is included in the account of the species and not in the account of the differences. And so the genus is not predicated *per se* of the difference, unless by mediation of its species, and therefore this is *per se*: 'the rational, insofar as it is a human being, is an animal'.

Further, this is clear from the Philosopher in *Topics* VI, who says that if the genus were predicated *per se* of the difference, one animal would be many animals. And this is clear since if what is rational *per se* is a rational animal, then through the account of being rational it is an animal, and since the account of being rational is included in the account of being a human being, a human being will be one animal through being rational. Further, a human being is one animal because being an animal falls into its definition, and being an animal does not fall into the definition of a human being through the account of being rational. A human being will be an animal, then, not through being rational, and so it will be another animal and thus one animal will be two animals.[9]

This is also a proof that being an animal does not fall into the definition of being a human being through being rational, since it would then be a nugatory definition.

Further, if being an animal were to fall into the definition of being a human being through the account of being rational, then that to which the account of being rational did not fit would also not be fit by the account of being an animal. Therefore, since the account of being rational does not fit with being a donkey, the account of being an animal would also not fit with it, which is false.

---

[9] Simon is concerned here about a nugatory definition of human being if the genus, animal, is predicated of the difference, rational. Since a human being is by definition a 'rational animal', then if 'rational' also includes 'animal' in its own definition, the definition of 'human' would be redundant, since 'animal' would be included twice.

⟨**Questio 19**⟩

De secundo arguitur.

1.1 Et arguitur quod hec non sit per se: 'homo est homo', quia 'per se; significat illud cui adiungitur esse causam alterius; denotans ergo hominem esse hominem per se denotat aliquid esse causam sui ipsius; sed hoc est impossibile; ergo etc.

1.2 Item, si hec sit per se, aut erit primo modo aut secundo aut quarto. Non primo, quia homo non est definitio hominis, nec pars definitionis; non secundo, quia non est propria passio hominis; non quarto, quia homo non est effectus hominis.

2 Oppositum arguitur: nulla propositio est verior illa in qua idem predicatur de se; sed ita est hic: 'homo est homo'; ergo est verissima, et ita erit ista per se.

3 Aliqui dixerunt quod ista est per se, quia tunc est propositio per se quando subiectum est causa quare predicatum ei inest; sed ita est hic: 'homo est homo', quia homo est homo per formam hominis que est in homine, et ita est per se.

Ista conclusio vera est, quia ista est per se: | 'homo est homo', sed non est vera ut ipsi intellexerunt, quia forma hominis est aliud ab homine, quia ⟨homo⟩ includit materiam cum forma; sed homo per nihil aliud ab homine est homo; ergo homo per formam hominis non est homo. Quod homo per nihil aliud ab homine sit homo probatio, quia per Philosophum[a] V *Metaphysice* dicitur quod, licet multe sunt cause hominis, sic⟨ut⟩ gressibile et bipes et talia, tamen homo est homo per illud quod est homo.

Item, nec hoc significat ista propositio 'homo est homo', scilicet quod homo sit homo per formam suam, quia relatio refert[i] illud quod precedit, sed

---

[a] Aristoteles, *Met.*, V, 18, 1022 a 32–34

---

3 alterius] alicuius M   20 quod] quia M   24 hoc] homo M

Question 19.

Next it is asked whether this is *per se*, 'The human being is a human being'.[10]

1.1 Concerning the second question, it is argued that 'A human being is a human being' is not *per se*, since '*per se*' signifies that that to which it is adjoined is a cause of another. One who indicates that a human being is a human being *per se* indicates that something is a cause of *itself*. But this is impossible. Therefore etc.

1.2 Furthermore, if this is *per se*, either it will be so in the first mode, or the second, or the fourth. It is not *per se* in the first mode, since 'human being' is not the definition of a human being, nor part of the definition. It is not *per se* in the second mode, because 'human being' is not a proper attribute of a human being. It is not *per se* in the fourth mode, because a human being is not the effect of a human being.

2 On the other hand, it is argued that no proposition is more true than one in which the same is predicated of the same. But 'a human being is a human being' is like this, therefore it is most true, and so it will be *per se*.

3 Some have said that this is *per se* since we have a *per se* proposition when the subject is the cause on account of which the predicate is in it. But 'A human being is a human being' is like this, since a human being is a human being through the form of human being, which is in the human being, and so the sentence is *per se*.

This conclusion is true, for 'a human being is a human being' is *per se*, but it is not true in the way that those who have so argued have understood it to be, since the form of a human being is distinct from a human being, for a human being includes matter with the form. Still, through nothing other than the human being is it a human being. Therefore a human being is not a human being through the form of human being. That a human being is a human being through nothing other than the human being is proved because, according to the Philosopher in *Metaphysics* IV, it is said that although there are many causes of a human being, such as being sociable, bipedal and the like, still a human being is a human being through that which is human being.

Further, neither does this proposition, 'a human being is a human being', signify this, namely, that a human being is a human being through his

---

[10] Cf. q. 18, l. 1–4.

non solum forma hominis precedit, et ideo cum dicitur sic: 'homo est homo per se', li 'per se' non refert tantum formam hominis sed totum compositum.

Unde dicendum quod homo per illud quod est est homo, et hoc ⟨non⟩ est 'per se' positive, ita quod habeat aliquam causam sui, sed privative, ita quod non habet ali⟨qu⟩am causam.

Unde considerandum quod quedam propositiones sunt ita ⟨per se⟩ vere quod rationes terminorum sunt omnibus note, ut 'de quolibet etc.', quedam sunt per se vere quarum rationes terminorum non sunt omnibus hominibus note, ut sunt iste 'homo est homo' et huiusmodi, tamen scienti rationes terminorum est per se manifesta. Probatio, quia illa propositio est per se, scilicet privative, cuius veritas dependet ex significato subiecti et predicati solum; sed talis est ista 'homo est homo'; ergo etc.

Et ad hoc attendens Philosophus[a] dixit fine VII *Metaphysice* quod querens quare homo est homo nihil querit, quia qui sic querit, idem querit et supponit; homo enim est homo, quia ipsum est ipsum. Quare dices tu quod non contingit sic querere, quia qui querit propter quid, scit quia est et si est, sed non contingit scire quia homo est homo et si homo est homo, et ⟨ideo⟩ propter quid homo est homo. Unde respondendum est sic querenti quod homo est homo quia est homo et huius est esse indivisibile.

Item, probatio quod iste qui negat istam 'homo est homo' non poterit predicare vere aliquid de homine, quia primum de quolibet genere est causa omnium posteriorum, et si primum in quolibet genere non est verum nec aliquod posteriorum; sed primum in genere predicationis est predicatio de se, cum sit simplicissima; qui ergo negat predicationem eiusdem de se negat omnem predicationem.

Sed tu dices: si ita sit in genere complexorum quod predicatio eiusdem de se est prima, ergo ista non est prima: 'de quolibet esse vel non esse'. Dico quod hoc est verum, immo ista prior 'ens est ens', et secundo 'non ens est non ens', et tertio 'de quolibet esse vel non esse', quarto 'de quolibet affirmatio vel negatio'.

[a] Aristoteles, *Met.*, VII, 17, 1041 a 14–24

27 li per] licet M   43 sed] et M   51 eiusdem] *iter.* M

form, for a relation refers to that which precedes it. But it is not only the form of human being that precedes, and therefore when it is said 'a human being is a human being *per se*', '*per se*' does not have reference only to the form of human being but to the whole composite.

Hence it must be said that a human being is a human being through that which it is, and this is not *per se* positively, so that it has some cause of itself, but privatively, so that it has no cause.

Thus it should be considered that some propositions are true *per se* in such a way that the accounts of the terms are known to all, for instance, that 'concerning everything it either is or is not'. There are some *per se* true propositions of which the accounts of the terms are not known to all, for instance, these: 'A human being is a human being', and the like. Nonetheless, to one who knows the accounts of the terms, it is clear that it is *per se*. The proof is that a proposition is privatively *per se* whose truth depends on the signified subject and predicate alone. But 'A human being is a human being' is like this; therefore etc.

Noting this, the Philosopher said at the end of *Metaphysics* VII that one who seeks out why a human being is a human being seeks nothing, since whoever seeks in this way seeks and presupposes the same thing. For a human being is a human being because it itself is itself. Hence you will say that it is not possible to seek in this way, since whoever seeks because-of-what-it-is knows that-it-is-so and if-it-is. But it is not possible to know that a human being is a human being and if a human being is a human being, and ⟨thereby know⟩ because-of-what a human being is a human being. Whence it must be answered thus to one who asks because-of-what a human being is a human being: because it is a human being, and its being is indivisible.

Furthermore, it is proved that one who denies 'A human being is a human being' cannot predicate anything of a human being truly, since the first in every genus is the cause of everything posterior ⟨in that genus⟩. And if the first in any genus is not true then neither will anything posterior be. But the first in the genus of predication is self-predication, since this is the simplest. So whoever denies predication of the same thing of itself denies every predication.

But you will say that if it is so in the genus of propositions that predication of the same thing of itself is first, then this is not the first: 'concerning everything, it either is or it is not'. I reply that that is true. Instead, this is prior, 'a being is a being'; in second place comes 'a non-being is a non-being', in third place, 'concerning everything, it either is or it is not';

Ad rationes.

Ad 1.1 Ad primam, dico quod 'per se' prout intelligitur positive denotat illud cui adiungitur esse causam alterius, tamen cum 'per se' intelligitur privative denotat non esse aliam causam illius cui adiungitur, et i[s]ta sumitur hic.

Ad 1.2 Ad aliam rationem, dico quod ista est per se quarto modo. [quod] Est per se quarto modo quod est per se manifestum, per Philosophum[a] V *Metaphysice*, ubi reducit omnes modos ad quattuor. Est etiam per se primo modo, quia tunc est per se primo modo quando ratio predicati est ratio subiecti vel ratio predicati includitur in ratione subiecti; unde, quia in proposito ratio predicati est ratio subiecti, ideo hec est per se. Iterum est in quarto modo, quia huius non est aliqua causa.

⟨Questio 20⟩

⟨De tertio arguitur.⟩

1.1 De tertio arguitur quod sic, hec sit per se 'animal est homo', quia quod per naturam suam et non per aliquod additum est homo per se est homo; sed animal est huiusmodi; ergo etc. Maior patet, quia 'per se' dicit idem quod 'non per aliud'; nunc autem animal est homo per rationale, quod non est aliud ab eo.

1.2 Item, si animal non per se est homo, ergo per accidens: aut ergo quia subiectum accidit predicato, aut e converso, aut utrumque alteri. Non primo modo, nec secundo, quia eadem natura importatur per utrumque, nec tertio modo, quia quod vere est ⟨res⟩ nulli accidit.

1.3 Item, 'homo est animal' convertitur in istam 'animal est homo'; sed hec est per se: 'homo est animal'; ergo etc. Et confirmatur ratio, quia sicut vera non convertitur nisi in veram, ita per se vera nisi in per se veram.

[a] Aristoteles, *Met.*, V, 18, 1022 a 25–36

68 aliqua] alia M

and in fourth place 'concerning everything, either its affirmation or its negation'.

In response to the arguments.

Ad 1.1 To the first I reply that *'per se'* as it is understood positively denotes that to which being a cause of another is adjoined; but when *'per se'* is understood privatively it denotes that there is no other cause of that to which it is adjoined, and this is how it is taken here.

Ad 1.2 To the other I reply that it is *per se* in the fourth mode. That is *per se* in the fourth mode that is clear *per se*, according to the Philosopher in *Metaphysics* V, where he reduces every mode to the fourth. It is also *per se* in the first mode, since it is *per se* in the first when the account of the predicate is the account of the subject, or the account of the predicate is included in the account of the subject. So, since the account of the predicate in the case at hand is the account of the subject, this is *per se*. Again, it is *per se* in the fourth mode since it has no cause.

Question 20.

Next it is asked whether 'The animal is a human being' is per se.

1.1 Concerning the third question, it is argued that this is *per se*, 'the animal is a human being', for what is a human being through its own nature, and not through anything else added to it, is a human being *per se*. But the animal is of this sort. Therefore etc. The major premise is obvious since *'per se'* indicates the same thing as 'not through another'. But the animal is a human being though being rational, since it is not other than that.

1.2 Furthermore, if an animal is not a human being *per se*, then it is so accidentally. Therefore this would be either because the subject is accidental to the predicate, or else conversely, or else each would be accidental to the other. And it is not in the first or the second way, since the same nature is conveyed through both, nor is it in in the third way, since what truly is a thing is not the accident of anything.

1.3 Furthermore, 'the human being is an animal' is converted to this, 'the animal is a human being'. But this is *per se*, 'the human being is an animal'. Therefore etc. And the argument is confirmed since just as what is true only converts to what is true, so what is *per se* true converts only to what is *per se* true.

2 Oppositum arguitur: 'per se' supponit 'de omni', si ergo hec sit per se: 'animal est homo', hec erit per se: 'omne animal est homo'; sed hoc est falsum; ergo et primum.

3 Quidam dicunt quod considerando ad ea ad que metaphysicus considerat, hec est per se, tamen considerando ut logicus, cum animal sit homo per rationale et rationale addit aliquid rationis super animal, ideo est per accidens.

Sed ista opinio falsa est, quia si esset per se, esset primo modo, quia non secundo, quia homo non est propria passio animalis, nec quarto modo, quia illius perseitatis est aliqua causa. Sed probatio quod non primo modo, quia tunc est propositio per se primo modo quando predicatum dicit quod quid est subiecti vel partem quod quid est; sed homo non est tale respectu animalis.

Item, falsum dicunt in hoc quod dicunt quod metaphysicus in perseitate non considerat rationes, | quia aliter esset hec per se: 'rationale est animal', quod ipse negat.   M 102va

Alii dicunt quod ista 'animal est homo' est per se perseitate predicati, non tamen perseitate subiecti; 'per se' enim dicit causam, et ideo 'per se' dicit perseitatem subiecti vel predicati. Si dicat perseitatem predicati, hec est per se, quia animal per hoc quod est homo est homo; si subiecti, falsa est, quia tunc significat quod animal secundum quod animal est homo.

Sed hec opinio est falsa, quia nulla propositio videtur esse per se nisi perseitate subiecti. Et hoc declaratur, quia propositio solum dicitur esse ⟨per se⟩ perseitate illius cuius per se est conditio; sed per se solum est conditio subiecti; ergo etc. Probatio minoris, quia ex dictis eorum, quia 'per se' dicit causam, ergo solum invenitur in eo in quo est causa inherentie; sed causa inherentie est solum in subiecto; ergo etc.

Item, si ista 'animal est homo' sit per se perseitate predicati, tunc significat quod animal per hominem est homo; sed hoc non significat ista propositio 'animal per se est homo', quia relatio solum refert illud quod precedit, sed li 'animal' solum precedit; ergo etc.

---

23 aliqua] alia M   25 homo] animal M   26 animalis] hominis M   32 subiecti] *iter.* M

2 On the other hand, it is argued, *'per se'* applies to every case. So if this is *per se*, 'the animal is a human being', then this will be *per se*, 'every animal is a human being'. But the second is false, therefore so is the first.

3 Some say that considering it in the respects in which a metaphysician does, it is *per se*. But considering it as the logicians do, one would argue that since an animal is a human being through being a rational being, and being rational adds something pertaining to reason over and above 'animal', it is thus accidental.

But this opinion is false, since if it were *per se* it would be so in the first mode, since it is not *per se* in the second mode, given that human being is not a proper attribute of animal, nor is it *per se* in the fourth mode, since there is a cause of that *per se* existence. The proof that it is not in the first mode is that a proposition is *per se* in the first way, when the predicate indicates the essence of the subject or a part of its essence. But a human being is not such with respect to animal.

Furthermore, they are wrong in saying that the metaphysician does not consider accounts when it comes to being *per se*, because otherwise this would be *per se*, 'the rational being is an animal', which the metaphysician denies.

Others say that this proposition, 'the animal is a human being', is *per se* by the *per se* status of the subject. For *'per se'* indicates a cause, and therefore *'per se'* indicates the *per se* status of the subject or the predicate. If it indicates the *per se* status of the predicate, this is *per se* since the animal, insofar as it is a human being, is a human being. If it indicates the *per se* status of the subject it is false, since in that case it signifies that an animal considered as such is a human being.

But this opinion is false, for no proposition seems to be *per se* except by the *per se* status of the subject. To explain this: that proposition alone is said to be *per se* by the *per se* status of that of which *per se* is a condition, but *per se* is only a condition of the subject, therefore etc. Proof of the minor premise: according to their claims, since *per se* indicates a cause, it is found only in that in which there is a cause of inherence, but a cause of inherence is only to be found in the subject. Therefore etc.

Furthermore, if this, 'the animal is a human being', is *per se* by the *per se* status of the predicate, then it signifies that the animal, through human being, is a human being. But this does not signify this proposition, 'the animal is *per se* a human being', since a relation refers only to that which precedes it, and only 'animal' precedes it; therefore etc.

Ideo dicendum est quod ista est per accidens: 'animal est homo', quia perseitas non solum attenditur penes rem sed penes rationem. Tunc licet eadem res importetur nomine animalis et hominis, tamen in natura animalis non reperitur aliquid determinate ratione cuius animal sit homo, sed accidit animali quod sit homo. Sed intelligendum quod diversimode sunt iste per accidens: 'animal est homo' et 'rationale est homo' et 'homo est albus', quia hec 'homo est albus' ⟨est⟩ per accidens quia res accidit rei, et hec est per accidens 'rationale est homo' quia ratio predicati accidit rationi subiecti et e converso.

Sed tu dices: estne hec 'animal est homo' reducibilis ad aliam per se? Dico quod sic sub disiunctione ad suum oppositum, ut 'animal est homo aut non homo', secundum quod enim dicit negationem in genere et non extra genus, et est per se secundo ⟨modo⟩.

Ad rationes.

Ad 1.1 Ad primam, dico ad maiorem quod insufficienter arguit, quia non solum ad perseitatem requiritur identitas rei sed rationis; unde illud quod per naturam suam et non per aliquid additum est homo, realiter est homo, non tamen per se.

Ad 1.2 Ad aliam, dico quod est per accidens. Ulterius dico quod predicatum accidit subiecto, quia ratio predicati accidit rationi subiecti, sed res non accidit rei; eadem enim est res in utroque, sed hoc non sufficit ad perseitatem.

Ad 1.3 Ad aliam, concedo maiorem, non tamen sequitur quod, licet una sit vera per se, quod convertatur in veram per se, quod patet in hac: 'omne grammaticum de necessitate est homo'.

---

48 ratione] rationis M   51 hec] hic M   56 enim] non M   61 aliquid] aliquod M
69 grammaticum] grammaticus M

Therefore it must be said that this is accidental, 'the animal is a human being', since being *per se* is introduced not solely by the thing but also by the account. Then, although the same thing is conveyed by the names 'animal' and 'human being', still in the nature of an animal there is not found anything of a determinate account by which an animal is a human being. Instead, it is accidental to an animal that it is a human being. But it must be understood that these are accidental in different ways: 'the animal is a human being' and 'the rational thing is a human being' and 'the human being is white'. For 'the human being is white' is accidental because a thing is accidental to a thing; and 'the rational thing is a human being' is accidental because the account of the predicate is accidental to the account of the subject and conversely.

But, you will say, is not this, 'the animal is a human being', reducible to another *per se*? I reply that it is, considered in disjunction with its opposite, as, for instance, in 'the animal is a human being or is not a human being'. For inasmuch as this indicates a negation within the genus and not outside the genus, it is also *per se* in the second mode.

In response to the arguments.

Ad 1.1 In response to the first I reply to the major premise that it is insufficiently argued, for not only is identity of the thing required for *per se* existence, but also identity of the account. Whence that which is a human being by its nature, rather than through something added to it, is in reality a human being, but still is not so *per se*.

Ad 1.2 To the other I reply that it is accidental. Further, I maintain that the predicate is accidental to the subject since the account of the predicate is accidental to the account of the subject. But a thing is not accidental to a thing, for it is the same thing in each, but this does not suffice for being *per se*.

Ad 1.3 As to the other, I grant the major premise, but it does not follow that, granting that a proposition is true *per se*, it may be converted into a true proposition *per se*. This is obvious in the proposition, 'Every grammarian is of necessity a human being'.

⟨Questio 21⟩

Consequenter queritur utrum hec sit per se: 'animal est rationale'.

1 Et arguitur quod non, quia si sit per se, aut erit primo modo, aut secundo, aut quarto. Non primo, quia nec rationale nec irrationale dicunt quod quid est animal⟨is⟩; nec secundo, quia non sunt proprie passiones animalis; non quarto, quia quartus modus est quando predicatum comparatur ad subiectum sicut effectus ad causam, sed sic non est in proposito; ergo etc.

2 Oppositum arguitur: per se dicit privationem causalitatis; sed istius non est alia causa; ergo etc.

3 Dicendum est quod ista est per se: 'animal est rationale vel irrationale', quia, sicut dicit Avicenna,[a] eadem est comparatio generis ad differentias et subiecti ad proprias passiones; sicut enim non cuiuscumque subiecti sunt quecumque passiones, sic nec cuiuscumque generis sunt quecumque differentie. Nunc autem animal per se dividitur per rationale et irrationale, et iste differentie non ⟨sunt⟩ nate dividere aliquod aliud genus. Ergo arguo sic: que sunt eiusdem communitatis cum alio, et sunt ei appropriate, insunt eo secundo modo; sed rationale et irrationale sunt huiusmodi respectu animalis; ergo etc. Sed tamen rationale et irrationale non sunt proprie passiones respectu animalis, sed quia propria passio predicata de materiali habet rationem informantis ipsum genus subiectum, ideo omnis predicatio per se formalis de materiali habet reduci ad secundum modum dicendi per se. Cum ergo differentie habeant rationem informantis ipsum genus, ideo ista perseitas habet reduci ad perseitatem secundo modo. Tamen ista non est per se: 'animal est rationale', quia rationale non universaliter inest omni animali, sicut nec par inest universaliter numero, sed hoc totum sub disiunctione 'par vel impar'; eodem modo ex parte ista animal est rationale vel irrationale.

Ad 1 Ad rationem, dico quod est per se secundo modo, nec tamen predicatum comparatur ad subiectum sicut propria passio, sed quia predicatum eo modo comparatur ad subiectum sicut forma ad materiam, ideo aliqualiter et magis improprie dicitur per se secundo modo 'animal est rationale vel irrationale' quam ille in quibus predicantur proprie passiones.

[a] Avicenna, *Sufficientia* (1508, f. 7vb)

---

20 materiali] animali M   24 animal] non *add. sed exp.* M   29 subiectum] cum *add. sed del.* M

Question 21.

Next it is asked whether this is *per se*, 'The animal is rational'.

1 It is argued that it is not, since if it is *per se* it will either be so in the first mode, or the second, or the fourth. It is not *per se* in the first mode, since neither 'rational' nor 'nonrational' indicate the essence of the animal. Nor is it *per se* in the second mode, since these are not proper attributes of animal. Nor in the fourth, since the fourth mode occurs when the predicate is related to the subject as effect to cause; but it is not thus in the case at hand. Therefore etc.

2 On the other hand, it is argued that *'per se'* indicates privation of causality, but there is no other cause of this, therefore etc.

3 It should be maintained that this is *per se*, 'An animal is rational or nonrational', since, as Avicenna says, the relation of the genus to its differences, and of the subject to its proper attributes, is the same; for just as not every attribute belongs to every subject, so neither does every difference belong to every genus. Now animal is divided *per se* into 'rational and nonrational', and these differences are not suited to divide any other genus. Therefore I argue thus: those things which are of the same generality with another and are made proper to it inhere in it in the second mode, but rational and nonrational are like this with regard to animal, therefore etc. But still 'rational' and 'nonrational' are not proper attributes with regard to animal. But since a proper attribute predicated of what is material has the account of what informs the subject genus, every *per se* formal predication of what is material must be reduced to the second mode of speaking *per se*. Since, then, the differences account for the account of the thing informing the genus, this *per se* status must be reduced to the second mode of being *per se*. But this is not *per se*, 'the animal is rational', since 'rational' does not inhere universally in every animal, just as 'even' does not inhere universally in number. But the totality of number is contained under this disjunction, 'even or odd', and in the same way, partwise, the animal is rational or nonrational.

Ad 1 In response to the argument I reply that it is *per se* in the second mode: not because the predicate is related to the subject as a proper attribute, but because the predicate is related to the subject in the same way that form is related to matter. So, speaking somewhat more loosely than when proper attributes are predicated, 'the animal is rational or nonrational' is also said to be *per se* in the second mode.

⟨Questio 22⟩

Consequenter queritur utrum hec sit per se: 'homo est animal', homine non existente.

1.1 Et videtur quod non, quia in essentialiter ordinatis, si non sit primum, nec aliquod posteriorum; nunc autem ista ordinem habent: ⟨compositio⟩ in re, compositio in intellectu et compositio in sermone, quare si non sit compositio in re, nec in aliquo posteriorum; sed cum homo non sit, non est vera compositio in re animalis ad hominem, quia nec homo in re existit, quare nec compositio in intellectu nec in sermone; quare etc.

1.2 Item, quod corrumpitur quantum ad suam essentiam, corrumpitur quantum ad | omne predicatum essentiale sibi inherens; sed nullo homine existente homo corrumpitur quantum ad suam essentiam ⟨et⟩ substantiam, quare quantum ad omne predicatum essentiale sibi inherens, quare quantum ad esse animal; quare etc. M 102vb

1.3 Item, quod corruptum est essentialiter non manet idem neque secundum nomen neque secundum rationem definitionis; sed homine non existente, homo corrumpitur essentialiter, quare nec manet idem ⟨secundum nomen⟩ nec secundum definitionem, quare nec 'homo' nec definitio hominis nec pars definitionis predicatur de ipso; sed animal est pars definitionis hominis; quare ipso corrupto nec predicatur animal de ipso vere nec per se. Maior manifesta est per Commentatorem[a] secundo *De anima*, ubi dicit quod generatio et corruptio sunt tales transmutationes secundum quas, si aliquid transmutatur, non manet idem secundum nomen et secundum rationem.

1.4 Item, nullum mortuum est animal per se; sed homine non existente homo est mortuus; ergo etc.

1.5 Item, ens de non ente non predicatur per se; sed homine non existente 'homo' non significat rem extra animam, sed 'animal' est nomen illius quod est extra animam, quia potest esse asinus; ergo homine non existente hec erit falsa: 'homo est animal'.

---

[a] Averroes, *Commentarium magnum in Aristotelis De anima libros*, II, 5, comm. 57 (ed. Crawford, p. 216)

---

7 existit] -tit *dub.* M

Question 22.

Next it is asked whether this proposition, 'a human being is an animal', is true *per se* when no human being exists.

1.1 It seems it is not since, in things ordered essentially to one another, if there is no first then neither is there anything posterior. In fact, however, composition in a thing, composition in the intellect and composition in speech have an order so that if there is no composition in a thing then neither is there composition in anything posterior. But since there is no human being, there is no true composition in reality of animal with human being, since no human being exists in reality, so that there is no composition in the intellect either, nor in speech. Therefore etc.

1.2 Furthermore, what is destroyed as regards its essence is destroyed as regards every essential predicate inhering in it. But if no human being exists, then human being is destroyed as regards its essence and its substance. Hence human being is destroyed as regards every essential predicate inhering in it, and hence it is destroyed as regards its being an animal. Therefore etc.

1.3 Furthermore, what is corrupted essentially does not remain the same, neither according to its name nor according to the account of its definition. But if no human being exists then human being is destroyed essentially, so that it remains the same neither in name nor in definition, hence neither 'human being' nor the definition of a human being nor part of its definition is predicated of it. But 'animal' is part of the definition of human being; therefore if human being is destroyed, 'animal' is not predicated of it truly nor *per se*. The major premise is obvious from the Commentator on *De anima* II, where he says that generation and corruption are changes such that if anything is changed in that way, the same thing does not remain according to name and according to its account.

1.4 Furthermore, nothing dead is an animal *per se*, but if no human being exists then the human being is dead; therefore etc.

1.5 Furthermore, being is not predicated of non-being *per se*. But if no human being exists, 'human being' does not signify a thing outside the soul. But 'animal' is a name of that which is outside the soul, since there can be a donkey. Therefore if no human being exists, this proposition 'human being is an animal' will be false.

2.1 Oppositum arguitur sic: quocumque casu contingente semper definitio et pars definitionis per se predicatur de definito; sed animal est pars definitionis hominis; ergo etc.

2.2 Item, quandocumque aliqua duo significant idem essentialiter et solum differunt sicut determinatum et indeterminatum, unum essentialiter predicatur de altero, quia talis est predicatio superioris de inferiori; sed homo et animal idem substantialiter significant et solum differunt in modo determinati et indeterminati; quare animal de homine etc.

2.3 Item, quandocumque aliquid est causa per se et sufficiens alicuius, quocumque alio posito vel remoto quod non tollit causam non tollit effectum; sed hominem significare animal est causa sufficiens quare homo est animal, quia Philosophus[a] dicit IV *Metaphysice* quod homo est animal quia hoc significat; ergo qui non tollit hominem significare animal non tollit hominem esse animal. Sed quocumque casu contingente homo semper significat animal. Quare etc. Et hoc confirmatur, quia vox non cadit a suo significato.

2.4 Item, nullo homine existente homo est homo per se; et homo est animal; ergo nullo homine existente homo est animal per se.

3.1 Aliqui ponunt quod in causatis esse in effectu non differt realiter ab esse, sed solum in modo intelligendi, et tamen dicunt quod ista est per se: 'homo est animal', nullo homine existente. Ratio autem eorum est ista, quia homo ex hoc quod homo nec est nec non est in effectu, quia ⟨si⟩ ex hoc quod homo esset in effectu, non posset non esse. Item, nec secundum quod homo est non ens, quia prohibitum est non esse [n]umquam ⟨fieri; sed⟩ fieri potest homo; ergo nec secundum se est nec secundum se non est. Et hec ⟨est⟩ ratio Avicenne.[b] Ex hoc arguitur: accidentale non transmutat essentiale; sed essentialis est inherentia animalis ad hominem; ergo quamvis ponitur hominem ⟨non esse⟩ in effectu, hec erit per se: 'homo est animal'.

---

[a] Aristoteles, *Met.*, IV, 4, 1006 a 30–33   [b] Avicenna latinus, *De philosophia prima vel scientia divina*, V, 1 (ed. Van Riet, p. 227)

2.1 On the other hand, it is argued that in every case that occurs, always both the definition and each part of the definition is predicated *per se* of what is defined. But 'animal' is part of the definition of a human being. Therefore etc.

2.2 Furthermore, whenever any two things signify the same essentially, and only differ as determinate and indeterminate, one is predicated essentially of the other, for predicating the superior of the inferior is like this. But 'human being' and 'animal' signify the same with respect to substance and differ only in the way determinate and indeterminate things do. Therefore 'animal' is predicated of human being etc.

2.3 Furthermore, whenever anything is the *per se* and sufficient cause of something, whatever else might be posited or removed, what does not take away the cause does not take away the effect. But 'human being' signifying animal is the sufficient cause for why a human being is an animal, for the Philosopher says in *Metaphysics* IV that a human being is an animal because this is what it signifies. Therefore whoever does not take away 'human being' signifying animal does not take away a human being's being animal. But whatever case occurs, 'human being' always signifies animal. Therefore etc. And this is confirmed since an utterance does not fall away from what it signifies.

2.4 Furthermore, when no human being exists, a human being is a human being *per se*, and a human being is an animal, therefore, when no human being exists, a human being is an animal *per se*.

3.1 Some posit that, in things that are caused, existing in actual effect does not differ in reality from existing, but differs only in the manner in which it is understood. Yet they also say that this is *per se*, 'a human being is an animal', when no human being exists. The reason for these claims is that a human being, from the fact of its being human, neither exists nor fails to exist in actual effect. For if a human being exists in actual effect from the fact of its being human, then it could not fail to exist. Furthermore, neither could it be a non-being considered as it is human, for it is forbidden that non-being should ever come to be; but it is possible for a human being to come to be. Therefore it neither exists in itself nor does it not exist in itself. And this is Avicenna's argument. From this it is argued as follows: The accidental does not change the essential. But the inherence of animal in human being is essential. Therefore even though human being is posited not to be in effect, this will be *per se*, 'a human being is an animal'.

Sed aliquis diceret: tu dicis quod esse et non esse accidunt essentie rei. Ego dico quod esse in effectu est de essentia rei, et hoc videbitur in secundo huius. Sed dicentes aliam positionem dicunt quod non, quia non contingit aliquid intelligere perfecte et ignorare quod est de essentia eius. Sed rosa non existente possum intelligere idem [non] essentialiter quod possum ipsa existente; secundum Algazelem[a] enim possum intelligere quaternarium, ignorare tamen si est. Ergo esse non est de essentia rei.

Sed tu dices: si nullus homo sit, essentia hominis aliquod esse habet, ad minus quidditativum, et tale est verum esse extra animam; esse verum extra animam est esse in effectu. Ad hoc posset dici quod esse quidditativum hominis est esse verum extra animam, verum est pro tanto, quia non dependet ab anima. Unde ulterius dico quod esse verum extra animam non est esse in effectu, quia esse verum extra animam debetur rei absolute, sed esse in effectu non, sed prout est terminus generationis.

Sed tu queres ulterius: si ita sit quod homo non sit animal per esse in effectu, ⟨cum⟩ non invenitur nisi triplex esse, scilicet esse in effectu et esse intellectum et significatum, ergo homo erit animal per aliquod istorum; et ista accidunt essentie hominis; ergo homo erit animal per aliquod accidens. Dico quod diminute arguis in sumendo esse; est enim aliud esse, scilicet quidditativum verum extra animam, et per illud homo est animal.

Ad rationes potest dici sustinendo istam positionem.

Ad 1.1 Ad maiorem, vera est. Ad minorem, 'ista sunt essentialiter ordinata etc.', conceditur totum. Et cum dicitur 'nullo homine etc.', dico quod falsum est, et tamen hoc est verum quantum ad existere in effectu, immo vera est compositio quantum ad esse reale | quidditativum per quod sibi animal inest; per ⟨es⟩se enim quidditativum animal inest per [es]se homini, quod sufficit.

M 103ra

Ad 1.2 Ad aliud, conceditur maior. Ad minorem, dico quod nullo homine existente, homo [non] corrumpitur quantum ad esse in effectu, quod acci-

---

[a] Al Ghazali, *Logica* (ed. Lohr, p. 247)

---

69 quia non] *inv.* M   75 aliquod] aliquid M   80 sunt] fuit M   87 ad] effe *add. sed del.* M

But someone might say, 'You maintain that being and non-being are accidents of the essence of a thing'. But I hold that being in effect is of the essence of a thing, as will appear in the discussion of the second book of the *Posterior Analytics*. Still, those holding to a different position deny this, for it is not possible for someone to understand perfectly and be ignorant of what its essence is. But when no rose exists I can understand the same thing non-essentially that I can understand when a rose exists. For, according to Alghazali, I can understand a notebook and yet not know if one exists, and therefore being is not of the essence of a thing.

But you will reply that if no human being exists, the essence of a human being nonetheless has some being, at least quidditative, and that sort of being is true being outside the soul. But being outside the soul is being in actual effect. To this it can be answered that the quidditative being of a human being is a true being outside the soul—true in this way because it does not depend on the soul. Hence I further maintain that true being outside the soul is not being in effect, for true being outside the soul needs to belong to the thing without qualification, whereas being in actual effect does not, but only insofar as it is the product of generation.

But you will ask next, if it is the case that a human being is not an animal through being in effect, then since there is no sort of being to be found except these three—being in effect, being understood and being signified—therefore a human being will be an animal through one of these. But these are accidental to the essence of a human being. Therefore a human being will be an animal through some accident. I reply that you leave something out in listing the sorts of being, for there is another being, namely true quidditative being outside the soul, and through this a human being is an animal.

A response to the arguments can be made by sustaining this position.

Ad 1.1 To the major premise, it is true. To the minor, 'these are essentially ordered', etc., this is entirely granted. And when it is said that 'when no human being' etc., I reply that this is false, and still this is true as far as existence in actual effect is concerned. Indeed it is a true composition as far as a real quidditative being is concerned, through which animal inheres in it. For animal, through quidditative being, inheres *per se* in a human being, which is enough.

Ad 1.2 In response to the other argument, the major premise is granted. To the minor premise I reply that, when no human being exists, human

dit homini per rationem Avicenne;[a] non corrumpitur ergo homo quantum ad suum esse quidditativum.

Et tu dices: illud tollitur per corruptionem quod acquiritur per generationem; sed hoc erat substantia rei; quare etc. Ad illam maiorem, conceditur. Ad minorem, cum dicitur 'per generationem acquiritur substantia rei etc.', dico quod non absolute acquiritur substantia rei per generationem, sed substantia que existit in effectu, ut quando generatur Cesar generatur substantia Cesaris existens.

Sed tu dices: si absolute substantia generetur, habetur propositum, si ⟨per⟩ accidens tunc esset alteratio. Dico quod nec istud nec illud, sed utrumque, substantia scilicet per respectum ad accidens. Nec iste respectus dicit aliquid additum nisi secundum rationem, quia generato homine albo generatur album non absolute nec per se, quia sic sibi non competit. Sic dico quod substantia generatur per respectum ad accidentia, non absolute, etc.

Ad 1.3 Ad aliam rationem, concedo maiorem. Ad minorem, dico quod non corrumpitur essentialiter quantum ad essentiam absolute, sed ut erat terminus generationis, et hoc modo illud est singulare cui competit generatio et corruptio, et propter hoc verum est quod homo corruptus est quantum ad essentiam ut erat terminus generationis [ut erat] singularium. Nec oportet tunc querere ubi est essentia; male enim queritur, quia essentia nec loco nec tempore mensuratur.

Ad 1.4 Ad aliud, dico quod 'mortuum' potest dicere privationem vite, et sic est maior vera, vel potest dicere privationem operationis consequentis vitam, et quoad hoc mortuum potest esse homo quantum ad essentiam.

Ad 1.5 Ad aliud, concedo maiorem. Ad minorem, dico quod nullo homine existente 'homo' non est nomen illius quod non est extra animam, immo dico quod nullo homine existente homo habet verum esse extra animam quod non dependet ab anima, et eius esse est verum reale quidditativum, sed hoc non est esse in effectu.

---

[a] Avicenna latinus, *De philosophia prima vel scientia divina*, V, 1 (ed. Van Riet, p. 227)

---

99 aliquid]   aliquod° M

being is destroyed with respect to the being in effect that is accidental to a human being, as in Avicenna's argument. Therefore human being is not corrupted with respect to its quidditative being.

You will answer, 'that is taken away through corruption which was acquired through generation, but this was the substance of a thing; therefore etc.'. In response to this I grant the major premise. To the minor, when it is said 'through generation is acquired the substance of a thing' etc., I reply that the substance of a thing is not acquired without qualification through generation, but that substance which exists in effect is acquired. For instance, when Caesar comes into being the existing substance of Caesar comes into being.

But you will reply, 'if substance comes to be without qualification, we have what was proposed, whereas if it is accidental then there is alteration'. I hold that neither one nor the other is so, but both of them: the substances, that is, in relation to an accident. Nor does that relation imply anything added except a relation of reason. For when a human being has come to be white, white comes to be neither absolutely nor *per se*, for it is not compared to itself in those ways. Thus I maintain that substance is generated through a relation to accidents, and not without qualification, etc.

Ad 1.3 In reply to the other argument I grant the major premise. To the minor I answer that it is not corrupted essentially in regard to its essence without qualification, but only as it was the end result of generation. And in this way that is a singular to which generation and corruption apply. And because of this it is true that a human being is destroyed with regard to his essence as it was a product of a generation of singular things. Nor is it necessary then to ask where its essence is. This is a bad question since its essence is measured neither by place nor by time.

Ad 1.4 To the other I reply that 'dead' can indicate the privation of life, and taking it in this way the major premise is true, or it can indicate the privation of the operation following on life. And a human being can be dead in this second way with respect to his essence.

Ad 1.5 In response to the other, I grant the major premise. To the minor premise I reply that when no human being exists, 'human being' is not a name of that which is not outside the soul. Indeed, I hold that when no human being exists human being has a true being outside the soul that does not depend on the soul, and its being is true real quidditative being. But this is not being in effect.

3.2 Potest etiam sustineri quod non sit per se. Et ratio huius est, quia 'per se' dicit causam, nunc autem ita est quod in unoquoque genere destructa causa destruitur effectus, nunc autem essentia hominis est causa quare homo est animal, ergo destructa ista essentia 'homo est animal' erit falsa; sed destructis singularibus destruitur ista essentia; ergo etc. Et hoc confirmatur per dictum Aristotelis[a] in fine VII *Metaphysice*, ubi dicit quod ⟨si⟩ trigonus esse⟨t⟩ mutabilis, dubium esset utrum haberet tres. Eodem modo possumus dicere in proposito.

Item, quando aliqua est propria dispositio immediate consequens essentiam alicuius, destructa illa dispositione destruitur et essentia quam consequitur, sicut patet de calefacere respectu ignis; sed esse in effectu est dispositio consequens essentiam hominis et cuiuslibet alterius causati; ergo destructo esse in effectu destruitur essentia hominis; et per essentiam hominis homo est animal; ergo etc.

Item, homo et animal sunt in uno predicamento; ens autem extra animam est quod dividitur in decem predicamenta et essentialiter predicatur de quolibet animali, ergo predicatum de homine et ipsum ens sunt vere extra animam; ergo posito quod nullus homo sit, si hec sit falsa 'homo est ens extra animam', hec erit falsa 'homo est animal'.

Ad rationes.

Ad 2.1 Ad primam, dico quod definitio et pars definitionis per se predicantur de definito, supposita autem essentia definiti, qui autem destruit essentiam definiti definitionem dicit ei non inesse; nunc autem qui ponit hominem non esse destruit essentiam hominis; ergo etc.

Ad 2.2 Ad aliam rationem, dico quod ad predicationem per se non sufficit quod idem significent quia significare accidit essentie, sed ⟨oportet⟩ quod essentialiter sint idem; nunc autem destructo homine homo et animal non sunt essentialiter idem; ideo etc.[b]

---

[a] Aristoteles, *Met.*, VII, 15, 1039 b 27–1040 a 7   [b] *deficit responsio ad* 2.3

---

121 ergo]   et M   ‖ ista]   causa *add. sed del.* M   ‖ est²]   erit M   ‖ animal²]   hec *add. sed forte canc.* M

3.2 ⟨An alternative reply⟩ One can also uphold the view that 'a human being is an animal' is not *per se*, and the reason for this is that '*per se*' indicates cause. Now in every genus, when the cause is destroyed the effect is destroyed. But the essence of human being is the cause for why a human being is an animal. Therefore if this essence is destroyed then this proposition, 'a human being is an animal', is false. But if the singulars are destroyed then this essence is destroyed, therefore etc. And this is strengthened by what Aristotle says at the end of *Metaphysics* VII, when he says that if a triangle were mutable, there would be a doubt as to whether it has three angles equal to two right angles. We could say the same about the question under discussion.

Further, when something is a proper disposition immediately consequent on something's essence, when that disposition is destroyed the essence that it is consequent on is also destroyed, as is clear of the power of heating in the case of fire. But being in effect is a disposition consequent on the essence of human being, as well as whatever else it causes. Therefore if being in effect is destroyed then the essence of human being is destroyed, and it is through the essence of human being that a human being is an animal. Therefore etc.

Further, human being and animal are in one category, but being outside the soul is divided into ten categories, and it is essentially predicated of every animal, therefore the predicate concerning human being and being itself are truly outside the soul. Therefore, supposing someone posits that there are no human beings, if this is false, 'a human being is a being outside the soul', then this will be false, 'a human being is an animal'.

In response to these arguments:

Ad 2.1 To the first I reply that a definition and a part of a definition are predicated *per se* of what is defined. But if we suppose that an essence is defined, then whoever destroys the essence of what is defined implies that its definition does not inhere in it. Now whoever posits that there are no human beings destroys the essence of human being, therefore etc.

Ad 2.2 In response to the next argument I reply that it does not suffice for *per se* predication that they signify the same, for signifying is an accident of essence, but it is required that they be the same in essence. Now if human being is destroyed then that human being and animal are not the same in essence. Therefore etc.

Ad 2.4 Ad aliam rationem, dicunt quod hec est per se: 'homo est homo', et tamen non ista: 'homo est animal', quia qui dicit 'homo est homo' dicit idem de se, qui autem dicit 'homo est animal', homine destructo destruitur causa quare animal ei inerat, et ideo hec non est per se: 'homo est animal', homine destructo. Qui etiam dicit 'homo est homo' dicit quod homo est illud quod essentialiter est homo. (Sed adhuc videtur quod illud non valet, quia vide qualiter sequitur 'homo est homo', quia essentia unius est essentia alterius, et tunc videtur quod sequitur 'ergo homo est animal' eodem modo.)

⟨Questio 23⟩

Consequenter queritur utrum hec sit per se: 'Sortes est homo'.

1.1 Et videtur quod non, quia sicut se habet questio ad questionem, sic propositio ad propositionem, quia omnis questio potest transformari in propositionem; sed questio 'quid est' presupponit questionem 'si est'; ergo etc. Ergo ista 'Sortes est homo' presupponit istam 'Sortes est'; sed ista non est per se: 'Sortes est', quia potest corrumpi; ergo nec ista est per se: 'Sortes est homo'.

1.2 Item, illud non inest alicui per se quod inest ei pro aliquo tempore determinato; sed 'Sortes' significat hominem pro tempore determinato.

1.3 Item, quando aliqua significant sub oppositis rationibus, unum non predicatur per se de alio; sed 'homo' et 'Sortes' sunt huiusmodi, quia 'homo' sub ratione indeterminati, 'Sortes' sub ratione determinati; ergo etc.

1.4 Item, orationes sunt vere quemadmodum et res, ergo de rebus transmutabilibus sunt propositiones habentes veritatem transmutabilem; ergo cum Sortes sit res transmutabilis, non erit ibi veritas nisi transmutabilis, et talis | non est per se; ergo etc.

2 Oppositum vult Philosophus[a] V *Metaphysice*, dicens quod hec est per se: 'Callias est homo', ergo et hec.

[a] Aristoteles, *Met.*, V, 18, 1022 a 25–27

---

9 sed] quia M

Ad 2.4 In response to the other argument, they say that this is *per se*, 'human being is human being', and yet that this is not per se: 'a human being is an animal'. For whoever says 'human being is human being' says the same thing of itself, but whoever says 'a human being is an animal', when human beings have been destroyed, destroys the cause for why animal inheres in it. And so this is not *per se*, 'a human being is an animal', when human being is destroyed. And whoever says 'human being is human being' says that a human being is that which is human essentially. (But yet it seems that this is not valid, for observe how 'human being is human being' follows because the essence of one is the essence of the other. It will then seem that 'a human being is an animal' follows in the same way.)

Question 23.

Next it is asked whether this is *per se*, 'Socrates is a human being'.

1.1 It seems it is not, since just as question is related to question, so proposition to proposition, for every question can be transformed into a proposition. But the question of what-it-is presupposes the question of if-it-is. Therefore etc. Therefore this, 'Socrates is a human being', presupposes this, 'Socrates is', but the latter is not *per se*, since he can be destroyed, therefore neither is this *per se*, 'Socrates is a human being'.

1.2 Furthermore, that does not inhere in something *per se* which inheres in it for some determinate time. But Socrates signifies a human being for a determinate time.

1.3 Furthermore, when terms signify under opposed accounts one is not predicated of the other *per se*. But 'human being' and 'Socrates' are like this, since 'human being' signifies under an indeterminate account, and 'Socrates' under a determinate account. Therefore etc.

1.4 Furthermore, expressions are true in just the way that the things are. Therefore, propositions concerning things that are changeable possess changeable truth. Therefore, since Socrates is a changeable thing, there will not be any truth there except the changeable sort, and that is not *per se*, therefore etc.

2 The Philosopher holds the opposite view. In *Metaphysics* V he says that this is *per se*, 'Callias is a human being', and therefore so too in the case at hand.

3 Aliqui dicunt quod ista propositio 'Sortes est homo' uno modo est per se, alio modo est per accidens. Dicunt enim quod Sortes et quodlibet aliud individuum in substantiis materialibus importat aliquid reale super speciem. Unde dicunt quod Sortes importat duo, scilicet naturam humanam et aliquid additum: ratione nature humane est per se, ratione superadditi non, sicut est de ista 'triangulus eneus est triangulus'. Et confirmant illud sic: quando aliquod predicatum inest alicui aggregato ita quod ratione unius inest ei per se et ratione alterius per accidens, uno modo est per se, alio modo per accidens; ideo, cum Sortes sit quid aggregatum ex substantia et accidente, ratione substantie sue est homo per se, ratione superadditi est homo secundum accidens.

Primum quod ipsi dicunt alicui non videtur habere veritatem, scilicet quod individuum aliquod reale addit supra speciem, quia illud reale additum quod ipsi ponunt non est nisi aliquod accidens per quod individuum est individuum; sed non contingit ponere aliquod accidens tale; ergo etc. Probatio minoris, quia illud quod secundum se non est individuum non est causa positiva individuationis; nunc autem nullum accidens est per se individuum, quia solum est per se individuum quod per se existit, sed nullum accidens existit per se.

Sed aliquis dicet quod maior non est necessario vera, quia illud quod secundum se non est ens est causa entitatis in alio, puta forma; a simili aliquid potest esse causa individuationis in alio quod secundum se non est individuum. Manifestum est quod postquam causa et causatum sunt per se differentie entis oportet quod sint entia. Unde licet forma non sit ens eo modo quo compositum est, cum ⟨...⟩ ipsum esse eodem modo quo compositum est illud, quod debet esse causa individuationis, oportet quod sit ipsa indivisio vel indivisionem habens. Non potest esse indivisio, cum indivisio sit sola privatio, et nullum accidens positivum est tale, ergo oportet quod illud quod sit causa individuationis sit habens indivisionem secundum se, et solum tale est individuum. Ergo quod est causa individuationis est individuum.

Item, illud non est rationabile quod ipsi dicunt quod individuum est ens secundum accidens, quia substantia est illud quod est ens secundum se et non secundum accidens, et ideo non videtur quod illud quod est maxime substantia sit ens secundum accidens.

---

22 aliquid] aliquod M 24 aliquid] aliquod M 43 differentie] *post corr.* M ‖ entia] entes M

3 Some say that this proposition, 'Socrates is a human being', is *per se* in one way, but in another way *per accidens*. For they say that Socrates and every other individual among material substances conveys something real over and above the species, whence they say that 'Socrates' conveys two things, namely human nature and something added. It is *per se* by reason of human nature, not by what is superadded, just as in this proposition, 'the bronze triangle is a triangle'. And they confirm it thus: when some predicate inheres in an aggregate in such a way that by reason of one it is in it per se, and by reason of another it is in it *per accidens*, then in one way it is *per se*, and in another *per accidens*. Therefore, since Socrates is an aggregate of substance and accident, by reason of his substance he is a human being *per se*, and by reason of what is superadded he is a human being accidentally.

The first thing they say, that is, that some real individual adds something over and above the species, does not seem to have truth to some, since that real addition that they posit is nothing except some accident through which the individual is an individual. But it is not possible to introduce some such accident, therefore etc. Proof of the minor premise: that which is not an individual in itself is not a positive cause of individuation. But no accident is *per se* individual, for that only is *per se* individual that exists *per se*, but no accident exists *per se*.

But someone might say that the major premise is not necessarily true, for what is not in itself a being is a cause of being in another, considered as a form. Similarly, something can be a cause of individuation in another which is not in itself an individual. It is obvious that, since the cause and the caused are *per se* differences of being, it is necessary that they are beings. Hence, although a form is not a being in the way that a composite is, since its very being...,[11] in the same way as a composite is that which must be the cause of individuation, it is necessary that it be undividedness itself or that it be that which has undividedness. It cannot be undividedness, since that is merely a privation, and no positive accident is such. Therefore it must be that the cause of individuation is something having undividedness in itself, and the only such thing is an individual. Therefore what is the cause of individuation is an individual.

Furthermore, what they say is not rational, that an individual is a being accidentally, for a substance is that which is a being in itself and not accidentally, and so it does not seem that that which is most a substance would be a being accidentally.

---

[11] There seems to be something missing from the surviving text at this point.

Dicendum est igitur quod in substantiis materialibus individuum non addit aliquid reale supra speciem, quia si addat, aut illud est de genere accidentis, quod non potest, quia omne accidens unde tale potest reperiri in pluribus et per nihil tale individuum est individuum. Ergo oportet quod sit substantia: aut ergo materia aut forma aut compositum; et hoc non potest esse materia nec forma, quia utrumque potest reperiri in pluribus; nec compositum, quia illud est de quo querimus. Nihil ergo reale addit individuum super speciem.

Sed tu queres: per quid ergo individuum est individuum? Dico quod illud per quod individuum est individuum est ipsa indivisio, nec habet causam aliam formalem positivam.

Ex istis, ad propositum dico quod ista est per se primo modo, et ita per se quod non per accidens. Et ratio huius est, quia illa propositio dicitur esse per se primo modo quando subiectum nihil addit reale supra predicatum, et ratio predicati includitur in ratione subiecti; sed utrumque reperitur in proposito; ideo etc.

Sed tu queres: si Sortes nihil reale addit, quid est illud quod addit? Dico quod solum addit habitudinem ad accidentia, et illud accidit substantie unde substantia est.

Et tu dices: Sortes addit illud accidens, et accidens quid reale est, ergo addit quid reale. Dico quod pro tanto dicitur esse accidens quia non inest substantie unde substantia est. Tamen non dicitur accidens sicut res accidentis, sed quedam est relatio, et multe sunt relationes que sunt de genere accidentis, sicut relatio quam habet materia ad formam, sed inest materie per substantiam suam, ut suppono nunc. Unde ille respectus quem habet ad accidentia et ad generationem, cuius est per se terminus, non est res accidentis, unde hec substantia formaliter est hec per hoc quod habet respectum ad accidentia. Et ad hoc Philosophus[a] dixit in VII *Metaphysice* quod quidquid generatur in alio, intendendo per hoc quod generatio non est alicuius nisi in quantum refertur ad aliud, ut ad accidentia et ad generationem, cuius est per se terminus.

Ad rationes.

Ad 1.1 Ad primam, dico quod questio 'quid est' presupponit questionem 'si est', non tamen si est in effectu. Sed illud quod debet habere quid est,

---

[a] Cf. Aristoteles, *Met.*, VII, 7, 1032 a 11–13

---

56 aliquid] aliquod M || supra] super^{ra} *sic* M

It should be held, then, that in the case of material substances the individual does not add anything real above the species, since if it adds such a thing it is perhaps of the genus of accident, but this cannot be, for every accident, as such, can be found in many, and through no such thing is an individual an individual. Therefore, it is necessary that it be a substance, and in that case it will be either the matter, the form or the composite. But it cannot be the matter or the form, since either of these can be found in many, nor can it be the composite, since that is what we are asking about. But then the individual adds nothing real over and above the species.

But you will ask, through what, then, is an individual an individual? I reply that an individual is an individual through this very undividedness, nor does it have any other positive formal cause.

Given all this, I reply to the case in question that 'Socrates is a human being' is *per se* in the first mode, and so *per se* because it is not *per accidens*. And the reason for this is that that proposition is said to be *per se* in the first mode in which the subject adds nothing real over and above the predicate, and the account of the predicate is included in the account of the subject. But both are found in the case at issue, therefore etc.

But you will ask: If Socrates adds nothing real, what is it that he adds? I reply that he adds only a disposition toward accidents and that is accidental to the substance as substance.

And you will reply that Socrates adds a certain accident, and an accident is something real, therefore it adds something real. I reply that that an accident is said to be such because it does not inhere in a substance as substance, but it is not said to be an accident as a thing belonging to the accident. Instead, it is a kind of relation, and there are many relations that are of the genus of accident, for instance, the relation matter has to form. But form inheres in matter through its substance, as I am now supposing, whence that relation which it has to accidents and to generation, of which it is the *per se* result, is not a thing belonging to an accident. Hence this substance is formally what it is because of this relation it has to accidents. And in this regard the Philosopher said in *Metaphysics* VII that everything is generated in another, meaning by this that generation does not belong to anything except insofar as that thing is referred to another, as for instance to accidents and to that generation of which it is the *per se* result.

In response to the arguments.

Ad 1.1 I reply to the first that the question of what-it-is presupposes the question of if-it-is, but not of if-it-is in effect. But that which needs to have

oportet quod ei non repugnet esse, unde esse in effectu non est propria passio alicuius, sed quod ex sua ratione non sit quid prohibitum esse. Sed cum dicit: '"Sortes est homo" dicit quid est Sortes', verum est, quare 'supponit Sortem esse', verum est, sic quod non sit prohibitum hoc esse, non tamen sic quod Sortes sit in effectu; et quia esse Sortis non est prohibitum, ideo potest habere quid est.

Ad 1.2 Ad aliam, nego minorem. Ad probationem, dico quod 'Sortes' significat hominem determinati temporis, hoc est, 'Sortes' significat hominem ut habet habitudinem ad tempus determinatum et locum determinatum, | et ista tamen non includuntur in sua ratione. Et ideo cum ista accidant essentie Sortis, et homo est de essentia Sortis, Sortes non erit homo per hec.

Ad 1.3 Ad tertiam rationem, concedo maiorem, minorem nego. Et ad probationem, cum dicitur '"homo" significat sub ratione indeterminati', hoc est: 'homo' significat aliquid quod est ex se ⟨in⟩determinatum, determinatum tamen ab alio; et cum homo non sit ex se determinatum, et Sortes ex se determinatum, sic dicendo 'Socrates est homo' ratio predicati cadit in ratione subiecti, quia non determinato ex se competit quod determinetur ex alio.

Ad 1.4 Ad quartam rationem, dico quod in Sorte est duo considerare: naturam hominis et respectum ad accidentia. Ratione nature humane non ⟨cadit⟩ sub generatione et corruptione, ratione respectus est transmutabilis, et sic cadit sub generatione et corruptione, et isto ultimo modo non potest 'homo' per se verificari de Sorte, primo tamen modo potest 'homo' de Sorte verificari per se.

⟨Questio 24⟩

Visis aliquibus circa primum modum dicendi per se, queratur de secundo modo dicendi per se. Et primo queratur utrum accidens sit ens per se.

1 Et arguitur quod sic, quia Philosophus[a] V *Metaphysice* dividit ens in ens per se et ens per accidens, et ens per se in decem predicamenta, et accidens est aliquod decem predicamentorum; ergo etc.

[a] Aristoteles, *Met.*, V, 7, 1017 a 8–30

---

98 includuntur] includitur M   99 homo¹] hec M

a what-it-is must be that to which being is not repugnant. Whence being in effect is not a proper attribute of anything. But what is prohibited is that it not exist *because* of its account. But when he says 'Socrates is a human being indicates what Socrates is', this is true, and so 'it presupposes that Socrates exists' is true. But it is true in such a way that Socrates is not prohibited from existing, rather than because Socrates exists in actual effect. And since the existence of Socrates is not prohibited, there can thus be a what-it-is to Socrates.

Ad 1.2 In response to the second argument, I deny the minor premise. I reply to its proof that 'Socrates' signifies a human being of a determinate time, that is, 'Socrates' signifies a human being as he has a disposition toward a determinate time and a determinate place. Yet this is not included in Socrates's account, and so, since these are accidental to the essence of Socrates, whereas human being is of the essence of Socrates, Socrates will not be a human being because of these.

Ad 1.3 In response to the third argument, I grant the major premise and deny the minor. And in response to the proof, when it is said that 'human being' signifies under an indeterminate account, that is, 'human being' signifies something that is of itself indeterminate but is determined by another, and since 'human being' is not determinate of itself, and 'Socrates' is determinate of himself, in saying 'Socrates is a human being', the account of the predicate falls under the account of the subject, for that it is not determinate of itself agrees with its being determined by another.

Ad 1.4 In response to the fourth argument, I reply that there are two things to consider in Socrates, the nature of a human being and his relation to accidents. With regard to his human nature he does not fall under generation and corruption, whereas with respect to his accidents it is he is mutable and so does fall under generation and destruction. And in this latter way 'human being' cannot be true *per se* of Socrates, but in the first way 'human being' can be true *per se* of Socrates.

Question 24.

With these things understood about the first mode of speaking *per se*, let us ask about the second mode of *per se*, and first let us ask whether an accident is a being *per se*.

1 It is argued that it is, for the Philosopher, in *Metaphysics* V, divides being into being *per se* and accidental being, and divides being *per se* into ten categories. Now, accidents are in one of the ten categories. Therefore etc.

2 Oppositum vult Philosophus dicens quod accidens non est ens per se, sicut musicum et album.

3 Manifestum quod in entibus est devenire ad aliquod primum quod habet esse ex ratione sua, quia si aliquid habet esse et non habet esse ex ratione sua, est in potentia ad esse, ergo oportet quod determinetur ad esse per aliquid aliud habens esse, quia nihil educit se ipsum de potentia ad actum. Quero tunc: determinatum ad esse, aut esse habet ex ratione sua, aut est in potentia ad esse. Si ex ratione sua habet esse, habetur propositum. Si non, determinatur ad esse per aliquid aliud; quero de illo alio, et sic in infinitum, aut erit devenire ad primum aliquod, et illud primum est causa omnium aliorum. Unde hec est per se: 'Deus est ens', et universaliter omne dictum de Deo quod non transit in materiam extrinsecam, cuiusmodi sunt 'Deus est movens' et huiusmodi; motus enim non est in substantia primi, sed motus est in re mota tamquam in subiecto et non in movente. Sed non querimus isto modo utrum accidens sit ens per se, sed querimus utrum sit per se aliquo alio modo.

Ad hoc est intelligendum quod aliquid est ens per se dupliciter: aut ita ens per se quod habet esse ita quod in suo esse non dependet ab anima, et sic dividitur ens per se in decem predicamenta; alio modo dicitur aliquid esse per se quod non indiget aliquo alio ad hoc quod existat, et talia sunt individua substantie, ut Sortes et huiusmodi, et ista que sic sunt entia per se sunt per se tertio modo. Unde Sortes subsistit per se tertio modo, quia subsistit per illud quod non est aliud ab essentia sua, sicut per presentiam forme in materia. Et sic ⟨accidentia⟩ non dicuntur esse in tertio modo dicendi per se.

Sed dicet aliquis: si Sortes existit per se ita quod in suo esse non indiget alio, ergo non est causatum. Dico ad hoc quod ista duo stant simul: quod aliquod sit causatum ab alio ita quod effective sit ab illo, et tamen formaliter ex se ipso, sicut homo formaliter est ens per se, effective tamen a prima causa, quod est universaliter omnium causa.

---

10 ex²] in M   13 determinatum] determinate M   15 aliquid] aliquod M   30 forme in materia] materie in forma M   30–31 et ... se] *post* predicamenta M   34–35 formaliter] forma est M

2 The Philosopher holds the opposite view, holding that an accident, for instance, musical and white, is not a being *per se*.

3 It is obvious that among beings one can arrive at some first which has being from its own account, for if anything has being and does not have being of its own account then it is in potentiality toward being, and therefore it is necessary that it be determined to being through something else that has being, since nothing brings itself from potentiality into actuality. So then I ask about some thing that is determined to be whether it has being from its account or else is in potentiality toward being? If it has being from its account, we have what was proposed. If not, it is determined to be through something else. So then I ask about that other, and so on to infinity, or else some first will be arrived at, and that first is the cause of all the others. Hence this is *per se*, 'God is a being', and so too for everything said about God that does not extend into extrinsic matter. Of the latter sort are 'God is a mover' and the like. For motion is not in the substance of the first being, but is instead in the thing moved as its subject rather than in the mover. But we do not ask in this way whether an accident is a being *per se*; instead, we ask whether it is *per se* in some other way.

In response to this, it should be understood that something is a being *per se* in two ways. In one way, being *per se* has being in such a way that it does not depend on the soul for its being, and being *per se* in this way is divided into the ten categories. In the other way, something is said to be *per se* that does not need another for it to exist, and individual substances are like this, such as, for instance, Socrates. And what is a *per se* being in this way is *per se* in the third mode. So Socrates subsists *per se* in the third mode, for he subsists through that which is not distinct from its essence—for instance, through the presence of form in matter. And so accidents are not said to be in the third mode of speaking *per se*.

But someone might say that if Socrates exists *per se* in such a way that he does not need another for his being, then he is not caused. I reply to this that these two are consistent: that something is caused by another so that it exists in actual effect from another, and yet it exists formally from itself. A human being, for example, is formally a being through itself (*per se*), but exists in actual effect from a first cause, which is the cause of all things universally.

Rationes procedunt viis suis.

Ad 1 Ratio enim ad primam partem ostendit quod accidens ita est per se quod in suo esse non dependet ab anima, et hoc concessum est.

Ad 2 Alia ratio probat quod accidens non est ens ita quod non indiget alio ad suum esse, et hoc est concessum.

⟨Questio 25⟩

Consequenter ⟨queritur⟩ utrum aliquod accidens per se insit subiecto.

1.1 Et arguitur quod non, quia per se et essentiale sunt idem, unde ubi nos habemus in nostra translatione 'per se' in alia habetur 'essentiale'; sed nullum accidens est essentiale; ergo etc.

1.2 Item, que per se insunt de necessitate insunt; sed accidentia non insunt de necessitate; ergo etc.

1.3 Item, diversitas modi inherendi sequitur diversitatem eorum que inherent, sed quando substantia inheret et quando accidens inheret diversum et diversum est inherens, ergo et diversus modus inherendi; sed substantia inheret per se; ergo et accidens per accidens.

2 Oppositum arguitur: conclusio demonstrationis est per se; sed in conclusione demonstrationis accidens inheret subiecto; ergo etc.

3 Dicendum est ad hoc quod aliquod accidens per se inheret subiecto. Ad cuius evidentiam considerandum est quod quedam accidentia sequuntur aggregatum secundum speciem, quedam secundum individuum. Accidentia que insunt secundum speciem insunt aggregato per essentiam propriam ipsius, non ita quod illa includuntur in ratione ipsius speciei, sed quod aggregatum per se sit causa eorum. Ex hoc sequuntur duo: primum est quod omne accidens proprium per se predicatur de suo subiecto; secundum est quod nullum accidens predicatur per se primo modo.

Probatio primi, quia aliquod predicatum | dicitur per se de subiecto quando in subiecto est causa predicati; sed universaliter subiectum est causa proprie passionis; ergo etc.

M 103vb

---

3 translatione] transmutatione M   7 sequitur] requiritur M   19 est] etiam M   ‖ accidens] per se *add. sed del.* M   22 universaliter] unu^{ter} M

The arguments go through in their own ways.

Ad 1 The argument to the first part shows that an accident is *per se* in such a way that in its being it does not depend on the soul, and this is granted.

Ad 2 The other argument proves that an accident is not a being in such a way that it does not need another for its being, and this too is granted.

Question 25.

Next it is asked whether any accident inheres in its subject *per se*.

1.1 It is argued that none do, since the *per se* and the essential are the same thing. Thus where we have '*per se*' in our translation, the other translation has 'essential'. But no accident is essential. Therefore etc.

1.2 Furthermore, what inheres in something *per se* inheres in it of necessity. But accidents do not inhere in things of necessity. Therefore etc.

1.3 Furthermore, a difference in ways of inhering follows from a difference in the things that inhere. But when substance inheres and when accident inheres there is a difference in what inheres. Therefore different sorts of inhering take place. But substance inheres *per se*, and therefore accident inheres *per accidens*.

2 On the other hand, it is argued that the conclusion of a demonstration is *per se*, but in the conclusion of a demonstration an accident inheres in a subject; therefore etc.

3 It should be replied to this that some accidents inhere in a subject *per se*. To make this evident it should be considered that some accidents follow an aggregate according to its species, whereas others do so according to the individual. Accidents that inhere in things according to species inhere in the aggregate through its proper essence, not so that they are included in the account of its species, but so that the aggregate is *per se* their cause. From this, two things follow. The first is that every proper accident is predicated *per se* of its subject. The second is that no accident is predicated *per se* in the first mode.

Proof of the first is that some predicate is said *per se* of its subject when the cause of the predicate is in the subject, but in every case the subject is the cause of its proper attribute; therefore etc.

Probatio secundi, quia in propositione per se primo modo predicatum non est extra intellectum subiecti; sed propria passio est extra intellectum sui subiecti, quia passio comparatur ad subiectum sicut effectus ad causam, sed effectus non est de intellectu cause; ergo etc. Quedam sunt accidentia que sequuntur aggregatum secundum individuum, et talia non habent causam propriam in subiecto, et ideo non insunt per se subiecto, sive sint separabilia sive inseparabilia. Unde hec non est per se: 'corvus est niger'.

Sed aliquis diceret: videtur quod accidentia inseparabilia insunt per se, quia insunt semper et de necessitate, ergo per se. Ad hoc respondet Expositor[a] quod non omnia accidentia inseparabilia insunt per se substantiis, quia si sunt talia accidentia que nec ponuntur in definitione substantiarum nec substantia in definitione eorum, non insunt per se. Sed ipse querit ulterius[b] qualiter sciemus que accidentia inseparabilia debent poni in definitione subiecti et que non. Et ipse respondet ad hoc quod si sint talia accidentia que habent propriam causam in subiecto, ita quod non possunt reperiri in aliquo subiecto [non] differente secundum speciem, talia ponuntur in definitione substantiarum. Modo ita est quod nigredo reperitur in aliis a corvo differentibus ab illo secundum speciem, et ideo nec nigredo ponitur in definitione corvi nec e converso.

Sed tu queres: cum sit aliquod accidens quod inest per se, estne aliquod accidens quod insit subiecto primo? Videtur enim quod non, quia quod primo inest alicui videtur esse de substantia eius; et accidens non est de essentia subiecti; ergo etc.

Item, intellectus componens sequitur intellectum apprehendentem; sed illud quod primo apprehendit intellectus est substantia rei, ergo quod primo componit cum re est de substantia eius; ergo nullum accidens primo componetur cum subiecto.

Ad hoc dicendum quod possumus considerare vel predicata de aliquo subiecto universaliter vel in genere: universaliter ita quod accipiamus omnia predicata essentialia et accidentalia, in genere ita quod accidentalia. Primo modo manifestum est quod nullum accidens inest subiecto primo, quia illud quod sic inest primo est substantia rei. Secundo modo accidens aliquod inest per se primo ipsi subiecto, cuiusmodi est risibile respectu

---

[a] Cf. Albertus Magnus, *De V universalibus*, tractatus de differentia, 2 (ed. Santos Noya, p. 82)
[b] Cf. Albertus Magnus, *De V universalibus*, tractatus de differentia, 3 (ed. Santos Noya, p. 85)

---

52 vel] ut M

Proof of the second is that in a proposition *per se* in the first mode the predicate is not outside the concept of its subject, but a proper attribute is outside the concept of its subject, for the attribute is to the subject as effect is to cause, but the effect is not part of the concept of the cause; therefore etc. Other accidents follow the aggregate according to the individual, and these do not have a proper cause in the subject, and therefore they do not inhere in the subject *per se*, whether they are separable or inseparable. Hence this is not *per se*, 'the crow is black'.

But someone may say that it seems inseparable accidents inhere in things *per se*, since they inhere in them always and of necessity, and therefore *per se*. To this the Expositor ⟨Albert the Great⟩ responds that not all inseparable accidents inhere in their subjects *per se*, since if there are such accidents that are neither included in the definition of the substance nor is the substance included in the definition of the accidents, then they are not in their subjects *per se*. But next this same person will ask how we know which inseparable accidents ought to be included in the definition of the subject and which ought not. The answer to this is that if they are accidents that have their proper cause in the subject, so that they cannot be found in any subject differing in species, then these are included in the definition of the substance. Now as it happens blackness is found in other things differing from crows in species, and therefore blackness is not included in the definition of crow, nor conversely.

But you ask whether, given that there is some accident which inheres in a thing *per se*, there is any accident which inheres in a subject primarily. For it seems that there is not, since what inheres primarily in a subject seems to be part of its substance, and an accident is not part of the essence of a subject; therefore etc.

Furthermore, the compounding intellect follows the apprehending intellect. But that which the intellect primarily apprehends is the substance of the thing, therefore what is primarily composed with the thing belongs to its substance. Therefore no accident is primarily composed with its subject.

To this it should be answered that we can consider a thing predicated of a subject in different ways, universally or in some genus, so that we take all essential and accidental predicates universally, but we take accidental predicates in some genus. In the first way, it is obvious that no accident inheres in its subject primarily, since that which inheres primarily is the substance of the thing. In the second way, some accident inheres *per*

hominis. Et hoc est necessarium, quia non est aliquod accidens ita communc quin insit alicui primo, quia sumpto aliquod subiectum cui inest accidens aut inest ei primo aut per aliud. Si primo, habetur propositum; si non, ergo inest illi per aliquod aliud, et quero sic de illo sicut prius, et ita erit processus in infinitum in predicatis accidentalibus, quod est contra Aristotelem V *Metaphysice*.[a]

Ad rationes patet.

Ad 1.1 Ad primam, concedo maiorem sumendo 'essentiale' large. Ad minorem dico quod aliquid esse essentiale est dupliciter: vel ita quod sit in sua ratione vel ita quod immediate causatur a subiecto et essentialiter; et isto ultimo modo insunt accidentia essentialiter subiecto.

Ad 1.2 Ad aliam rationem, nego minorem. Ad probationem dico quod hoc est intelligendum de accidentibus que consequuntur aggregatum secundum individuum; illa enim sunt corruptibilia et nulli insunt per se.

Ad 1.3 Ad aliam, dico quod ratio non concludit plus nisi quod modus inherendi accidentis non est per se eo modo quo modus inherendi substantie, et hoc concedo. Concedo etiam quod accidentia insunt per accidens, hoc est, non insunt eo modo per se quo illa insunt que sunt per se primo modo. Illa insunt per se primo modo quando predicatum cadit in definitione subiecti; secundus modus est quando subiectum est causa predicati, et ita est quando propria passio predicatur de proprio subiecto.

---

[a] Aristoteles, *Met.*, VII, 6, 1031 b 18–1032 a 5

---

61 accidentalibus] essentialibus M   77 ita] illa M

se and primarily in its subject. Of this sort is the ability to laugh, with respect to a human being. And this is necessary since there is no accident so common that it does not inhere primarily in something, since, if an accident inheres in a subject, either it inheres in it primarily or through another. If primarily then we have our thesis; if not, then it inheres in the subject through some other and I ask of this other as before. And so there will be an infinite regress in accidental predicates, which is opposed to Aristotle in *Metaphysics* V.

The replies to the arguments are clear.

Ad 1.1 In response to the first I grant the major premise, taking 'essential' in a broad sense. With regard to the minor premise, I hold that 'some essential being' can be taken in two ways, in one way as that which is in the thing's account, in the other as that immediately and essentially caused by the subject. And in this second sense accidents inhere essentially in the subject.

Ad 1.2 In response to the second I deny the minor premise. To prove it wrong, I hold that this is to be understood about accidents that follow an aggregate as it is an individual, for these accidents are corruptible, and inhere in nothing *per se*.

Ad 1.3 In response to the other, I reply that it does not conclude anything more than this, that the way in which the accident inheres is not *per se* in that way in which the substance's way of inhering is. I grant this. I also grant that accidents inhere in things *per accidens*, that is, they are not in things *per se* in the same way that those things are that are in things *per se* in the first mode. Those inhere in things *per se* in the first mode when the predicate falls into the definition of the subject. The second mode is when the subject is the cause of the predicate, and that is when a proper attribute is predicated of its proper subject.

⟨Questio 26⟩

Consequenter queritur circa illud, scilicet utrum duo accidentia possint primo inesse eidem subiecto, ut par vel impar numero.

1.1 Et arguitur quod non, quia sicut se habent duo subiecta ad unum accidens primum, sic duo accidentia ad unum subiectum primum; sed duo subiecta non possunt primum esse subiectum unius accidentis primi, quia tunc duo subiecta essent unum subiectum; quare duo accidentia non possunt primo inesse eidem subiecto.

1.2 Item, subiectum per suam essentiam est immediata causa accidentis; sed unius cause est unus effectus per se et primo; ergo etc.

1.3 Item, quod per superhabundantiam dicitur convenit uni soli;[a] sed primo inesse est quid dictum per superhabundantiam; ergo etc.

2 Oppositum ⟨patet⟩ per intentionem Philosophi[b] I *Metaphysice*. Pythagoras voluit quod illud quod primo inest alicui esset substantia eius. Arguit Philosophus contra eum, quia si illud quod primo inest alicui esset substantia eius, multa essent substantia unius; sed ista consequentia non valeret nisi plura accidentia immediate inessent alicui subiecto.

3 Aliqui volentes solvere questionem istam dicunt quod aliquid dicitur inesse alicui primo quia inest non mediante priori. Sed hoc potest esse non mediante priore subiecto, sicut est hec 'triangulus habet tres': habere tres inest primo | triangulo, quia non mediante priore subiecto, tamen inest sibi mediante priore predicato, quia per hoc quod duo anguli valent tres angulos extrinsecos sibi oppositos. Vel non mediante priore predicato, et hoc modo inest definitio definito, quia non mediante priore subiecto nec mediante priore predicato. Tunc cum queritur utrum duo accidentia etc., dicunt quod duo accidentia ⟨non⟩ possunt simul inesse eidem subiecto primo, ita quod non mediante priori predicato, quia si sic passio esset indemonstrabilis de subiecto, quia demonstratio est per causam. Sed si ita esset quod non esset aliquod predicatum essentiale quod prius diceretur de subiecto quam passio, non esset causa quare passio diceretur de subiecto; nec erit definitio substantie [nec qua aliud], et ita ⟨nec⟩ propria passio nec aliquod accidens probaretur de subiecto.

M 104ra

---

[a] Aristoteles, *Top.*, V, 5, 134 b 23–24   [b] Aristoteles, *Met.*, I, 5, 987 a 23–29

---

2 vel] np *add. sed del.* M   13 substantia] esse *add. sed del.* M   19 hec] hic M   30 erit] enim M   ‖ substantie] *ut v. s. l.* M

Question 26.

Next we ask whether two accidents can inhere primarily in the same subject, for instance, even or odd in number.

1.1 It is argued that they cannot, since as two subjects are related to one first accident, so are two accidents to one first subject. But two subjects cannot be the first subject of one first accident, for then the two subjects are one subject. Hence two accidents cannot inhere primarily in the same subject.

1.2 Furthermore, a subject is through its essence the immediate cause of an accident, but of one cause there is one *per se* first effect, therefore etc.

1.3 Furthermore, whatever is said through superabundance agrees with one thing only. But 'inhering primarily' is something said through superabundance, therefore etc.

2 The opposite view is clear from the meaning of the Philosopher in *Metaphysics* I. Pythagoras thought that what inheres in something primarily is its substance. The Philosopher argued against him, since if that which inheres primarily in something were its substance, there would be many substances of one subject. But this inference would not hold unless many accidents were to inhere immediately in some subject.

3 Some who wish to resolve this question say that something is said to inhere in something primarily because it does not inhere in it by the mediation of anything prior. But this can mean unmediated by a prior subject in the way this is, 'a triangle has three ⟨angles equal to two right angles⟩'. 'Having three ...' inheres in a triangle primarily since it is not mediated by a prior subject, but, still, it inheres in it by mediation of a prior predicate, since it inheres in it through the fact that the two angles are equal to the three exterior angles opposite to them. Alternatively, 'inhere primarily' can mean not mediated by a prior predicate. And in this way the definition is in the defined, since it is not mediated by a prior subject, nor by a prior predicate. Then, when it is asked whether two accidents etc., they say that two accidents cannot be primarily in the same subject at the same time in such a way that they are not mediated by a prior predicate. For if this could be the case then an attribute would be indemonstrable of its subject. For a demonstration is through a cause, but if it there were not some essential predicate that were said of a subject prior to the attribute, then there would be no cause for why an attribute is said of its subject. There will also not be a definition of the substance,

Isti verum dicunt, sed non dissolvunt. Certum enim est quod duo accidentia non insunt primo subiecto, non mediante priore predicato, sed si insunt primo insunt primo ita quod non mediante priori; sed hoc est
35 quod querimus, et ergo ad illud non respondent.

Et ideo dicendum quod secundum hunc modum non possunt duo accidentia inesse primo subiecto uni per unum et primo, sed per aliud et aliud. Et huius declaratio est, quia diversis actibus respondent diverse potentie, quia potentiam et actum oportet esse proportionata; sed si duo acciden-
40 tia insunt primo eidem subiecto, illa accidentia sunt duo diversi actus; oportet tunc quod in ipso subiecto sint due diverse potentie, ita quod per unam potentiam recipiat primo unum accidens et per aliam aliud.

Sed tu dices: ex dictis tuis sequitur quod uni subiecto inquantum unum non inest nisi unum accidens primo, quod videtur esse falsum, quia nu-
45 merus est unum et tamen par et impar insunt ei primo. Dico quod si ista duo accidentia par et impar insunt numero primo, non insunt ei nisi ut unum. Et quomodo unum? Unum genere, quia numerus est unum genere. Si tamen consideretur par et impar secundum proprias rationes ipsorum, sic considerata non insunt numero secundum quod est quid unum, sed
50 secundum quod diversificatur secundum diversas species, quod Lincolniensis[a] dicit, quod nihil prohibet unum et idem dici de pluribus, ita quod de quolibet illorum primo, sicut genus inest pluribus speciebus ita quod cuilibet illorum primo; sic de diversis accidentibus respectu subiecti. Si nos consideramus aliquod unum inesse duobus primo, non contingit nisi
55 altero istorum duorum modorum: vel quod illa plura considerantur ut unum, vel quod illud unum consideretur ut plura. Sic oportet considerare de genere altero istorum modorum respectu specierum ad hoc quod insit eis primo. Sic patet quod duo accidentia insunt eidem primo, sed per aliud et aliud, sicut in corpore animato; vivere enim inest ei per animam
60 primo, primo etiam inest ei esse mortale, et hoc est per materiam.

Ad 1.1, 1.2 et 1.3 Et per hoc apparet ad rationes.

[a] Robertus Lincolniensis, *In Posteriorum analyticorum libros*, I, 8 (ed. Rossi, pp. 119–22)

35 ergo] etiam M   56 oportet] cumd *add. sed del.* M

and so neither a proper attribute nor any accident would be proved of its subject.

These respondents speak truly, but they do not resolve the problem. For it is certain that two accidents do not inhere in a first subject unmediated by a prior predicate, but if they inhere in it primarily then they do so in such a way that they are not mediated by a prior. But this is what we seek, and to that, therefore, they do not respond.

And so it must be said that in this way there cannot be two accidents inhering primarily in one subject through one thing, primarily, but only through distinct things. And the explanation of this is that different potentialities correspond to different actualities, since a potentiality and actuality must be proportionate to one another. But if two accidents inhere primarily in the same subject, then those accidents are two different actualities. It is necessary, then, that in the subject there be two different potentialities, so that through one potentiality it receives the one accident primarily, and the other through the other.

But you will reply that from what you have said it follows that nothing inheres primarily in one subject insofar as it is one except one accident, which seems to be false, since one is a number and yet both even and odd inhere in it primarily.[12] I reply that if these two accidents, even and odd, inhere primarily in a number then they inhere in it only as it is one. And one in what way? One in genus, since number is one in genus. But if even and odd are considered according to their own proper accounts, then so considered they do not inhere in number as it is something one, but as it is diversified into different species. This is what Grosseteste says, namely that nothing prevents one and the same thing being said about many in such a way that it is said primarily of every one of them, as a genus inheres in many species in such a way that it is said primarily of each of them. So too of various accidents with respect to a subject. If we consider some one thing to inhere primarily in two things, this is possible only in one of these two ways: either because those many things are considered as one, or because that one thing is considered as many. Thus it is necessary to consider the second of these ways with respect to species inasmuch as they inhere primarily in those. Thus it is obvious that two accidents inhere primarily in the same subject, but through different causes, as in an animated body. For life inheres in it primarily through the soul, and being mortal also inheres in it primarily, but this is through its matter.

Ad 1.1, 1.2, and 1.3 This makes the reply to the arguments apparent.

---

[12] For Aristotle's discussion of the view that unity is both even and odd, see *Metaphysics* 986a.

⟨Questio 27⟩

Consequenter queritur utrum hec sit per se: 'numerus est par'.

1 Et videtur quod non, quia in propositione per se secundo ⟨modo⟩ oportet quod predicatum insit subiecto de necessitate; sed par non de necessitate inest numero; ergo nec per se.

2 Oppositum arguitur: accidentia insunt per se subiectis que accipiuntur in ratione illorum accidentium, sicut vult Philosophus[a] in capitulo isto; sed in definitione paris ponitur numerus, quia numerus par est numerus non habens medium; quare videtur quod hec sit per se secundo modo.

3 Dicendum quod ista non est per se: 'numerus est par', ista tamen est per se: 'numerus est par vel impar'. Similiter ista non est per se: 'linea est recta', ista tamen est per se: 'linea est recta vel curva'. Primum patet, quia ad hoc quod propositio sit per se secundo modo oportet quod predicatum insit subiecto de necessitate et universaliter; sed hoc predicatum par non de necessitate et semper et universaliter inest numero, nec rectum linee, quia non de necessitate linea est recta nec omnis linea est recta, et similiter ⟨nec⟩ de necessitate numerus est par nec omnis numerus est par; quare non erunt per se, cum per se et de omni idem sint, et que sunt per se sunt necessaria, ut ostensum est supra.

Sed dicet aliquis: ad propositionem per se non requiritur, ut videtur, nisi quod in subiecto sit causa predicati; 'per se' enim causam importat, quare sufficit ut in subiecto sit causa predicati vel e converso; nunc autem in linea invenitur sufficiens causa recti et in numero paris; quare etc.

Dicendum est ad hoc quod verum est quod ad propositionem per se secundo modo requiritur quod subiectum sit causa predicati sufficiens, et cum hoc quod subiectum sit causa precisa predicati. Et hoc patet ex precedentibus, quia ad propositionem per se secundo modo oportet quod predicatum et subiectum dicantur secundum convertentiam, hoc autem non esset nisi subiectum esset causa precisa predicati. Sic autem non est cum dicitur 'numerus est par', quia numerus non est causa ⟨precisa⟩ respectu huius quod dico par, quia est causa alterius. Tamen ista est per se secundo modo: 'numerus est par vel impar', quia ista est per se secundo

---

[a] Aristoteles, *AnPo*, I, 4, 73 a 37–73 b 2

7 paris] patris *ante corr.* M   8 medium] numerus M   27 subiectum] secundum M

Question 27.

Next it is asked whether this is *per se*, 'number is even'.

1 It seems it is not, since in a proposition *per se* in the second mode, it is necessary that the predicate inhere in the subject of necessity, but being even does not inhere in a number of necessity, therefore it is not *per se*.

2 On the other hand, it is argued that accidents inhere in subjects *per se* which are contained in the account of their accidents, as the Philosopher holds in this chapter. But in the definition of 'even', number is included, since an even number is a number not having a middle. Thus it seems that this is *per se* in the second mode.

3 It should be replied that this is not *per se*, 'number is even', but this is *per se*, 'a number is even or odd'. In the same way, this is not *per se*, 'a line is straight', but this is *per se*, 'a line is straight or curved'. The first is obvious, since for the proposition to be *per se* in the second mode it is necessary that the predicate inhere in the subject of necessity and universally. But this predicate, even, does not inhere in number of necessity and always and universally, nor does straight inhere in line, for a line is not necessarily straight, nor is every line straight. And in the same way number is not of necessity even, nor is every number even. Hence it will not be *per se*, for *per se* and *in every case* are the same, and what is *per se* is necessary, as was shown above.

But someone will reply that, as we have seen, nothing is required for a *per se* proposition except that in the subject there be the cause of the predicate. For *per se* conveys cause, so it suffices that the cause of the predicate be in the subject or vice versa. But in a line a sufficient cause of straight is found, and in a number, a sufficient cause of even. Therefore etc.

It is to be replied that it is true that it is required for a proposition *per se* in the second mode that the subject be the sufficient cause of the predicate, but also this is required, that the subject be the *precise* cause of the predicate. And this is obvious from the preceding, since for a proposition *per se* in the second mode it is necessary that subject and predicate be said convertibly, but this will not be unless the subject is the precise cause of the predicate. But this is not so when it is said, 'number is even', since number is not a precise cause with respect to this that I call 'even', for it is the cause of another ⟨predicate⟩ as well. But this is *per se* in the second

modo: 'numerus est numerus habens medium vel non habens medium', sicut ista: 'animal est homo vel non homo', ut ostensum est. (Sed 'omne habens medium in genere numerorum est par vel impar' non est per se.) Sed que per se insunt de necessitate insunt, quare ista 'numerus est | par vel impar', 'linea est recta vel curva' sunt per se secundo modo. Unde iste tres propositiones differunt: 'numerus est par', 'numerus medium non habens est par' et 'binarius est par', quia ista: 'numerus est par' nullo modo est per se; hec 'numerus medium non habens est par' est per se primo modo, quia subiectum est causa inherentie predicati ad ipsum; hec 'binarius est par' est per se non primo: non est primo, quia binarius ex hoc quod binarius non est par, sed ex hoc quod numerus medium non habens.

Ad rationes.

Ad 1 Ad primam, dico quod maior vera est, hoc supposito, quod illa accidentia universaliter insint subiecto; hec enim est una conditio requisita ad propositionem per se. Ad minorem, dico quod quamvis numerus ponatur in definitione paris, tamen par non inest universaliter numero, et ideo hec non est per se, 'numerus est par'. Et si vellemus aliter dicere, quod non sufficit ad propositionem per se quod predicatum includit subiectum in sua definitione, sed quod subiectum sit immediatum respectu predicati; item, quod ipsum subiectum sit causa efficiens et precisa ipsius et secundum quod ipsum est unum. Numerus non est causa paris per hoc quod numerus, sed per hoc quod est numerus medium non habens.

⟨Questio 28⟩

Consequenter queritur circa tertium modum utrum tertius modus sit modus inherendi vel essendi.

1.1 Et arguitur quod sit modus inherendi, quia ille modus per se est modus inherendi ubi 'per se' est determinatio inherentie; sed in tertio modo est 'per se' determinatio inherentie, quia ista est per se tertio modo: 'Sortes est per se', sed 'per se' est hic determinatio inherentie esse ad Sortem; ergo etc.

---

34 non] que M   45 supposito] suppono M   46 subiecto] subiecti M   51 sit] inn *add. sed del.* M

mode, 'a number is even or odd', since this is *per se* in the second mode, 'a number is a number having a middle or not having a middle', just as this is, 'an animal is a 'human being or not a human being', as has been shown. (But 'everything having a middle in the genus of numbers is even or uneven' is not *per se*.) But what inheres in something *per se* inheres in it of necessity, so these propositions, 'a number is even or uneven', 'a line is straight or curved', are *per se* in the second mode. Hence these three propositions differ: 'a number is even', 'a number not having a middle is even', and 'what is twofold is even'. For this, 'a number is even', is in no way *per se*. This, 'a number not having a middle is even', is *per se* in the first mode, since the subject is the cause of the inherence of the predicate. This proposition, 'what is twofold is even', is *per se*, but not in the first mode. It is not *per se* in the first mode because what is twofold is not even because it is twofold, but because it is a number not having a middle.

In response to the arguments.

Ad 1 To the first I reply that the major premise is true, supposing that these accidents inhere in the subject universally, for this is a condition required for a *per se* proposition. To the minor premise I reply that although number is included in the definition of even, still being even is not in number universally. And so this is not *per se*, 'number is even'. (⟨This is so⟩ even if we might wish to say otherwise: that it does not suffice for a *per se* proposition that the predicate include the subject in its definition, but it is also required that the subject be immediate to the predicate and, also, that the subject itself be the efficient cause and the precise cause of this predicate, and insofar as the subject is one thing. ⟨For⟩ a number is not the cause of being even because it is number, but because it is a number that does not have a middle.)

Question 28.

Next, it is asked concerning the third mode, whether it is a mode of inhering or of being.

1.1 It is argued that it is a mode of inhering, since that *per se* mode is a mode of inhering when '*per se*' is a determination of inherence. But in the third mode there is a *per se* determination of inherence, since this is *per se* in the third mode, 'Socrates is *per se*'. But '*per se*' here is a determination of the inherence of being to Socrates, therefore etc.

1.2 Item, Philosophus[a] in capitulo de per se solum numerat modos qui conferunt ad demonstrationem; sed illi qui conferunt ad demonstrationem sunt modi inherendi; ergo etc.

2 Oppositum arguitur: ille modus per se non est modus inherendi secundum quem modum aliquod incomplexum dicitur per se, quia ubicumque est inherentia est quedam complexio predicati cum subiecto; sed penes tertium modum aliquod incomplexum dicitur per se, puta Sortes et omnis prima substantia; ergo etc.

3 Lincolniensis[b] dicit quod omnes modi per se quos Philosophus dat reducuntur in tres modos, scilicet in modos essendi, inherendi et modum causandi. Primus modus et secundus sunt modi inherendi, quartus modus causandi, tertius essendi. Et in hoc convenit cum Themistio[c] super passum illum, qui dicit quod Philosophus non enuntiat illum tertium modum ut conferat ad demonstrationem, sed ut completus sit numerus per se. Sed planum est quod si esset modus inherendi, conferret ad demonstrationem, quare manifestum est quod iste tertius modus non est modus inherendi.

Item, quod non sit modus inherendi patet sic, quia secundum istum tertium modum substantia singularis dicitur per se; nunc autem substantie singularis non est inherere alii, nec etiam substantie universalis, quia nomine substantie importatur aliquid absolutum, et isti sunt modi per quos Philosophus distinguit substantiam ab accidente, scilicet quod substantia est per se ens, accidens autem est alii inherens; manifestum igitur quod iste tertius modus dicendi per se non est modus inherendi.

Et hoc est quod Philosophus[d] dicit in littera, quia secundum istum modum dicuntur aliqua per se que non dicuntur de subiecto aliquo, talia autem sunt sola singularia. Singularia autem sunt per se subsistentia, quia subsistunt per aliquid quod non est extra rationem eorum, tunc ad hoc quod hec sint non requiritur aliquid aliud cui accidunt, hec sunt igitur subsistentia per se. Ideo iste tertius modus non est modus inherendi, sed modus per se subsistendi.

Ad 1.1 Ad rationem primam, concedo maiorem. Nego minorem, quia per se tertio modo non dicit inherentiam predicati ad subiectum. Dico quod 'per se' potest determinare hoc verbum 'est' vel ipsam inherentiam.

---

[a] Aristoteles, *AnPo*, I, 4   [b] Robertus Lincolniensis, *In Posteriorum analyticorum libros*, I, 4 (ed. Rossi, pp. 111–12)   [c] Themistius, *Analyticorum posteriorium paraphrasis*, 4 (ed. O'Donnell, p. 255)   [d] Aristoteles, *AnPo*, I, 4, 73 b 6–10

---

9 sed ... demonstrationem] *i. m.* M   12 quia] sed M   16 dat] dicit M   31 non] *s. l.* M

1.2 Furthermore, the Philosopher in the chapter on *per se* enumerates only those modes that relate to demonstration. But those which relate to demonstration are modes of inhering. Therefore etc.

2 On the other hand, it is argued that that *per se* mode is not a mode of inhering according to which a term is called *per se*, because wherever there is inherence there is some sort of composition of a predicate with a subject. But in the third mode a term is called *per se*, for instance, Socrates, along with every other primary substance. Therefore etc.

3 Grosseteste says that all the modes of *per se* that the Philosopher sets out are reduced to three modes, namely to the modes of being, inhering, and causing. The first and second modes are modes of inhering, the fourth is a mode of causing, and the third is a mode of being. And in this he agrees with Themistius's commentary on that passage, who says that the Philosopher does not set out the third mode so as to apply to demonstration, but only to complete the number of what is *per se*. But it is clear that if it were a way of inhering it would apply to demonstration, so it is obvious that this third mode is not a mode of inhering.

Furthermore, that it is not a mode of inhering is obvious, since in this third mode a singular substance is called *per se*. But, in fact, for a singular substance there is no inherence in another, nor is there even for a universal substance, for the name 'substance' conveys something without qualification. And these are the ways in which the Philosopher distinguishes substance from accident, namely, that substance is a *per se* being, whereas accident is that which inheres in another. Therefore it is obvious that this third mode of saying *per se* is not a mode of inhering.

This is what the Philosopher says in the text, for according to this mode things are said *per se* which are not said of a subject, and singular things are the only things of this sort. But singulars are *per se* subsisting things, for they subsist through something not outside their account, and hence it is not required that there be anything other to which they are accidental. These are, therefore, *per se* subsisting things. Therefore this third mode is not a mode of inhering, but a way of *per se* subsisting.

Ad 1.1 In response to the argument, I grant the major premise, and deny the minor, since *per se* in the third mode does not indicate the inherence of a predicate to the subject. I reply that *per se* can determine either the

Si determinet inherentiam, dico quod hec 'Sortes est per se' est per se secundo modo, et est sensus: Sortes est per se, scilicet esse per se inest Sorti, supponendo quod esse in effectu insit ei per se secundo modo. Si autem determinet hoc verbum 'est', sic 'Sortes est per se' ⟨est per se⟩ tertio modo, et est sensus: 'Sortes est per se', id est solitarius.

Sed aliquis dicet: ex dicto tuo videtur quod 'per se' faciat propositionem modalem, quia 'per se' potest determinare inherentiam vel hoc verbum 'est' per te, et ita hec erit modalis: 'homo est animal per se' et talia. Dico quod hoc est verum, quia non solum isti modi 'necessarium' et 'impossibile' et huiusmodi faciunt propositiones modales, sed infiniti alii faciunt propositiones modales.

Item, omnes iste sunt modales: 'Sortes bene currit' vel 'cito', quia modus non est nisi determinatio inherentie predicati ad subiectum, et ideo omnis illa determinatio que potest determinare inherentiam predicati ad subiectum potest facere propositionem modalem; huiusmodi autem sunt infiniti, et hoc vult Armonius[a] et Simplicius super librum *Perihermeneias*.

⟨Questio 29⟩

Consequenter queritur utrum Sortes sit per se tertio modo; secundo utrum quartus modus sit modus inherendi.

1.1 De primo arguitur quod non, quia 'per se' dicit privationem causalitatis cuiuslibet alterius, illud | ergo est per se tertio modo, et solum, quod non habet causalitatem sui esse, quare causato repugnat per se hoc modo; sed Sortes habet causalitatem sui esse; ergo Sortes non est per se tertio modo.

M 104va

1.2 Item, nihil quod est ens secundum accidens est per se tertio modo; sed Sortes est ens secundum accidens, quia in esse coniunctum est accidentibus, et etiam secundum aliquos in ratione sua includit duo realia quorum unum accidit alii; ergo etc.

---

[a] Ammonius, *In Aristotelis De interpretatione commentarius* (ed. Verbeke, pp. 14–15)

---

53 iste] isti M   5 quare] quia M   6 sed] quia M

verb 'is' or inherence itself. If it determines inherence, I reply that this, 'Socrates is per se', is *per se* in the second mode, and the sense of 'Socrates is *per se*' is that being inheres in Socrates *per se*, supposing that being in effect inheres in him *per se* in the second mode. But if it determines the verb 'is' then 'Socrates is *per se*' is in the third mode, and the sense is that Socrates is *per se*, that is, by himself.

But someone will say, from what you have said it seems that *per se* produces a modal proposition, since, according to you, *per se* can determine either inherence or the verb 'is', and so this proposition will be modal, 'a human being is an animal *per se*', and those like it. I reply that this is true, for not only these modes—necessary, impossible and so on—make propositions be modal, but an infinite number of others also make propositions be modal.

Again, all of these are modal: 'Socrates runs well' or 'quickly', for a mode is nothing except a determination of the inherence of the predicate to the subject. And so every determination which can determine the inherence of a predicate in its subject can make a modal proposition. But things of this sort are infinite, and this is the meaning of Ammonius and Simplicius on the *De interpretatione*.

Question 29.

Next it is asked whether Socrates is *per se* in the third mode and, secondly ⟨q. 30⟩, whether the fourth mode is a mode of inhering.

1.1 Concerning the first, it is argued that it is not, since '*per se*' indicates the privation of all other causality. Therefore, that is *per se* in the third mode, and only that, which does not have causality for its being, and so being caused is contradictory to being *per se* in this mode. But Socrates has causality for his existence, and therefore Socrates is not *per se* in the third mode.

1.2 Furthermore, nothing that is a being accidentally is *per se* in the third mode. But Socrates is a being accidentally, since he is conjoined with accidents in being, and also since, according to some, two realities are included in his account, of which one is accidental to the other. Therefore etc.

2 Oppositum arguitur: omne quod est species vel individuum de genere substantie est per se tertio modo, substantia enim est que ⟨est⟩ primo et principaliter; sed Sortes est in genere substantie; ergo etc.

3 Ad hoc est intelligendum quod Sortes est per se tertio modo, quia omne quod existit per rationem suam [est] et non per aliquid quod est extra rationem suam est per se tertio modo; ⟨sed Sortes existit⟩ ex ratione sua; ideo etc.

Item, omne quod significat hoc aliquid est per se tertio modo, sicut vult Philosophus[a] in littera, quia illud est per se tertio modo quod non dicitur de aliquo, sicut album et musicum; sed Sortes est hoc aliquid, idem enim est individuum et hoc aliquid, sed dicitur individuum quia non divisum in plures partes, et hoc aliquid quia per se subsistit; unde Sortes per se subsistit, et omne tale est per se tertio modo.

Sed tu dices: illa sunt per se tertio modo que significant hoc aliquid et per se subsistunt; quare videtur quod universalia substantie non sint per se tertio modo, quia non subsistunt per se sed in singularibus nec significant hoc aliquid quia dicuntur de singularibus; et illa sunt per se tertio modo quecumque non dicuntur de subiecto aliquo, et cum universalia dicantur de subiecto aliquo, videtur quod etc.

Dicendum est ad presens quod universale dicitur dupliciter: universale in actu et universale in potentia. Universale in potentia est natura rei intellecta in potentia, et ista natura est una in uno et plures in pluribus, et non una in pluribus antequam intellecta sit, et ideo ipsa non est aliud quam natura singularium; immo si singularia per se subsistant, per ipsam subsistunt, et ideo ipsa sic considerata est res per se tertio modo dicendi 'per se'. Alio modo dicitur universale in actu, et tale est natura rei actu intellecta ut in ipsa fundatur ratio universalitatis, et ideo in universali in actu est duo considerare, scilicet naturam subiectam universalitati et ipsam universalitatem. Considerando naturam subiectam est per se tertio modo, quia per illam naturam sunt particularia illud quod sunt, immo nec singularia sunt per se tertio modo nisi quia sunt per illam. Considerando illam universalitatem non est per se tertio modo: sic est accidens, et ad hoc considerans Aristoteles[b] VII *Metaphysice* contra Platonem dixit quod universalia non sunt substantie, et hoc est verum considerando ad rationem

---

[a] Aristoteles, *AnPo*, I, 4, 73 b 6–8   [b] Aristoteles, *Met.*, VII, 13

14 genere substantie] *inv.* M   17 ex] in M   36 est] sunt M

2 On the other hand, it is argued that everything which is a species or individual of the genus of substance is *per se* in the third mode, for a substance is what exists primarily and principally. But Socrates is in the genus of substance. Therefore etc.

3 In reply, it should be understood that Socrates is *per se* in the third mode, since everything that exists through its account and not through anything outside its account is *per se* in the third mode. But Socrates exists through his account, therefore etc.

Further, everything that signifies some particular (*hoc aliquid*) is *per se* in the third mode, as the Philosopher holds in the text, for that is *per se* in the third mode which is not said of anything, as with white and educated. But Socrates is a particular thing, for being an individual and being some particular thing are the same. But he is called an individual because he is not divided into many parts, whereas he is called a particular thing because he subsists *per se*. Whence Socrates subsists *per se*, and every such being is *per se* in the third mode.

But you will say those are *per se* in the third mode that signify some particular thing and subsist *per se*, on account of which it seems that universal substances are not *per se* in the third mode, since they do not subsist *per se*, but rather in singulars. Nor do they signify some particular thing, for they are said of singulars and those are *per se* in the third mode which are not said of a subject. So since universals are said of a subject, it seems that they are not *per se* in the third mode.

It should be replied to this that 'universal' is said in two ways, universal in act and universal in potential. A universal in potential is the nature of a thing understood in potential, and this nature is one in one thing and many in many, and not one in many before it is understood, and so it is nothing other than the nature of the singulars. But indeed, if singulars subsist *per se* then they subsist through this ⟨nature⟩, and so it is, so considered, a thing *per se* in the third mode of speaking *per se*. In another way 'universal' is said in act, and such is the nature of a thing understood in act, so that the account of universality is founded on that nature, and so two things are considered in the universal in act, namely the nature subject to the universality and the universality itself. Considering the nature subject to the universality, it is *per se* in the third mode, since through that nature particulars are that which they are. Indeed, singulars are *per se* in the third mode only because they exist through their nature. Considering that universality, a universal is not *per se* in the third mode,

universalitatis. Et hoc est quod dicit Commentator[a] XII *Metaphysice*, quod universalia apud Aristotelem sunt in anima collecta a pluribus singularibus ab intellectu considerante et faciente ea ex similitudine intentionem unam, et ista intentio accidit rei, et quantum ad illud non sunt universalia per se tertio modo. Nec valet 'universalia sunt in singularibus, ergo non sunt per se tertio modo'; bene enim concluderet ratio, si universale esset in singularibus sicut accidens in subiecto; nunc autem non est ita, sed universale est in singularibus sicut natura in habente naturam.

Ad aliud, dico quod hoc non est proprie dictum: 'universale est in aliquo sicut in subiecto'; 'subiectum' enim de sua ratione nominat aliquod ens in actu, et per hoc distinguitur subiectum a materia per Commentatorem.[b] Si igitur aliquid adveniat subiecto, est accidens, cum ipsum sit actu. Et ideo, cum omne predicatum de subiecto ut subiectum est accidens, [et] propter hoc non est verum dicere quod universalia predicentur de singularibus ut de subiecto aliquo, cum universalia sint de substantia singularium. Apparet igitur quod Sortes sit per se tertio modo, et universaliter omnis substantia prima et secunda ⟨est⟩ per se tertio modo dicendi per se, sicut prius visum est.

Ad 1.1 Ad rationes in oppositum, cum arguitur '"per se" dicit privationem causalitatis', dico quod 'per se' loquendo proprie dicit privationem causalitatis, ⟨sed⟩ aliquando dicit privationem unius cause, aliquando plurium. Cum enim dicitur 'Deus est per se', 'per se' dicit privationem cuiuslibet cause, quia principium non est per aliquam aliam causam a se. Alio modo dicit privationem causalitatis materie, ⟨sicut⟩ cum dicitur 'intelligentia est per se', quia forma earum non est recepta in materia. Alio modo dicit privationem subiecti, et hoc modo substantia tam prima quam secunda est per se, quia non indiget alio ut subiecto ad hoc quod subsistat. Sic autem omnis substantia est per se tertio modo, quia non habet causam aliam sui esse in genere ⟨subiecti⟩ , ergo etc. Dicitur in minori 'Sortes habet aliam causam', verum est, efficientem et materialem, tamen non habet aliam causam in genere subiecti.

[a] Averroes, *Commentarium magnum in Metaphysicam Aristotelis*, XII, 1, comm. 4 (1562, f. 292D)
[b] Averroes, *Commentarium magnum in Metaphysicam Aristotelis*, VIII, 1, comm. 3 (1562, f. 210GH); cf. I, 8, comm. 17 (1562, f. 14K)

48 intentionem] intentionum M 51 concluderet] consideret M ‖ ratio] rationem *ante corr.* M 54 ad aliud] *Cf.* Q. 29, 3 54–55 universale ... subiecto] *iter.* M 62 secunda] secundo M

and so it is an accident. With this in mind, Aristotle, in *Metaphysics* VII, said against Plato that universals are not substances. And this is true, considering them with respect to the account of universality. And this is what the Commentator says, on *Metaphysics* XII, that universals according to Aristotle are collected in the soul from many singulars by the intellect that considers and makes them into one intention due to their likeness. And this intention is accidental to the thing and with respect to that universals are not *per se* in the third mode, nor does the inference hold, 'universals are in singulars, therefore they are not *per se* in the third mode'. For the argument would indeed be valid, if the universal were in singulars as an accident is in a subject. But it is not so; rather the universal is in singulars as a nature is in what has a nature.

In response to the other, I reply that this is not properly said: that a universal is in something as in a subject. For 'subject', by its account, names some being in act, and through this the subject is distinguished from matter by the Commentator. If, then, something attaches to the subject it is an accident, since the subject is already in act. And therefore since everything predicated of the subject as subject is an accident, still, it is not true to say on account of this that universals are predicated of singulars as of a subject, since universals belong to the substance of singulars. It is apparent, then, that Socrates is *per se* in the third mode, and in every case every substance, both primary and secondary, is *per se* in the third mode of speaking *per se*, just as was seen earlier.

Ad 1.1 In response to the arguments opposed to this, when it is argued that *per se* indicates the privation of causality, I reply that *per se*, strictly speaking, does indicate the privation of causality, but sometimes of one cause and sometimes of several. For when it is said that 'God is *per se*', *per se* indicates the privation of every cause, since that which is principle does not exist through any cause other than itself. In another way it indicates the privation of the causality of matter, as when it is said that the separate intelligences are *per se*, since their form is not received in matter. In still another way it indicates the privation of the subject, and in this way both primary and secondary substances are *per se*, since they do not need another as their subject to subsist. But in this way every substance is *per se* in the third mode, since it does not have any other as the cause of its being in the genus of subject, therefore etc. It is said in the minor premise, 'Socrates has another cause'. This is true for efficient and material causes, but he does not have another as cause in the genus of subject.

Ad 1.2 Ad aliam rationem potest dici dupliciter; ille enim qui sustinet quod individuum includat duo realia, dicet sic: aliquid dicitur esse dupliciter secundum accidens, aut quia in esse coniunctum [in]est accidentibus, aut quia non est nisi cum accidit alii. Unde illud quod est ens secundum accidens isto secundo modo non est per se tertio modo, illud tamen quod est ens secundum accidens quia in esse coniunctum est accidentibus est per se tertio modo. Et secundum illam opinionem dicendum quod nihil quod includit duo realia quorum unum accidit alii est ens per se tertio modo. Nunc autem ita est quod Sortes non includit duo realia quorum unum accidit alii, ideo Sortes, et quecumque | alia substantia, potest esse per se tertio modo. M 104vb

⟨Questio 30⟩

⟨De secundo arguitur.⟩

1.1 De secundo arguitur, scilicet quod quartus modus non sit modus inherendi, sic: causare non est per se inherere, immo ex hoc quod aliquid causat aliud per se concluditur quod unum non est aliud, quia causa et causatum differunt in essentia; cum ergo quartus modus, secundum Lincolniensem,[a] sit modus causandi, ergo quartus modus etc.

1.2 Item, ex quo 'per se' dicit causam, ubicumque est modus inherendi per se, oportet quod predicatum comparetur ad subiectum. Si ut causa, sic est primus modus; si ut causatum, sic est secundus. Non enim ponit nisi duos modos inherendi, ergo omnis modus inherendi erit primus vel secundus.

2 Oppositum arguitur: omnis modus qui cadit in demonstratione est modus inherendi, et propter hoc voluit Themistius[b] quod tertius modus non est modus inherendi, quia non cadit in demonstratione; sed quartus modus cadit in demonstratione; ergo etc.

---

[a] Robertus Lincolniensis, *In Posteriorum analyticorum libros*, I, 4 (ed. Rossi, pp. 111–12)
[b] Themistius, *Analyticorum posteriorium paraphrasis*, 4 (ed. O'Donnell, p. 255)

---

83 illam] aliam M 2 arguitur] scilicet quod tertius modus *add. sed del.* M 8 si ut] sicut M 9–10 non ... inherendi] *post* causandi M

Ad 1.2 In response to the other argument, one can reply in two ways. For he who holds that an individual includes two realities will say this, that something is said to exist accidentally in two ways, either because it is conjoined with accidents in being, or because it exists only when it is accidental to something else. Accordingly, that which is an accidental being in this second way is not *per se* in the third mode, but that which is an accidental being because it is conjoined with accidents in being is *per se* in the third mode. With respect to this opinion, it should be held that nothing that includes two realities, of which one is accidental to the other, is a being *per se* in the third mode. But Socrates does not include two realities of which one is accidental to the other. Therefore Socrates, and every other substance, can be *per se* in the third mode.

Question 30.

Whether the fourth mode is a mode of inhering.

1.1 Concerning the second question it is argued that the fourth mode is a mode of inhering. For to cause is not to inhere *per se*. Indeed, from the fact that something causes another *per se* it is concluded that the one is not the other, for the cause and what is caused differ in essence. Since, then, the fourth mode, according to Grosseteste, is a mode of causing, the fourth mode is therefore a way of inhering.

1.2 Furthermore, since '*per se*' refers to the cause, wherever there is a mode of inhering *per se* it is necessary that the predicate be related to the subject. If it relates as a cause then it is related in the first mode; if it relates as that which is caused then it is related in the second mode. Therefore, since he posits only two modes of inhering, every mode of inhering *per se* will be the first or the second mode.

2 On the other hand, it is argued that every mode that falls within a demonstration is a mode of inhering. And it is for this reason that Themistius holds that the third mode is not a mode of inhering, for it does not fall within a demonstration. But the fourth mode falls within a demonstration. Therefore etc.

3 Ad hoc dicendum est quod quartus modus est modus inherendi per se, quia ibi est modus inherendi per se ubi subiectum comparatur ad predicatum ut causa ad causatum. Sic in quarto modo comparatur subiectum ad predicatum sicut causa ad causatum, et hoc in ratione cause efficientis, ut interfectus interiit propter interfectionem et ipsius non est alia causa, propter quod manifestum est quod quartus modus est modus inherendi. Unde subiectum comparatur ad suam passionem in duplici genere cause, scilicet in genere cause materialis et in genere cause efficientis: secundum quod comparatur ad suam passionem in genere cause materialis est secundus modus, secundum quod comparatur subiectum ad suam passionem in genere cause efficientis est quartus modus. Unde subiectum prout comparatur in ratione cause efficientis est principium demonstrandi passionem de se ipso in genere cause materialis, quia subiectum per sua principia formalia est causa effectiva passionis, ideo definitio que aggregat principia formalia subiecti est medium ad demonstrandum passionem de subiecto, sicut patebit post. Et quantum ad propositum uterque modus est modus inherendi, sed modus inherendi primus cadit in premissis demonstrationis, sed secundus in conclusione.

Sed quo modo differt iste modus ab aliis modis inherendi? Aliqui dicunt quod iste quartus modus differt a primo et a secundo sicut universale a particulare. Et hoc confirmant per Aristotelem,[a] ubi dat modos de 'secundum quod': primus modus 'secundum quod' pertinet ad formam; secundus ad materiam; et addit ulterius, 'secundum quod' dicit habitudinem cuiuslibet cause; tunc est quartus modus dicendi 'secundum quod'. Eodem modo dicunt de 'per se' quoad quartum modum dicendi per se.

Sed isti ad pauca considerabant. Verum est quod 'secundum quod' primo modo pertinet ad formam, secundo modo ad materiam, sed quod ipsi dicunt, quod in quarto modo 'secundum quod' dicit habitudinem cuiuscumque cause, ⟨falsum est⟩. Dicit enim Aristoteles[b] quod illa est per se quarto modo cuius non est aliqua causa, cuiusmodi est ista 'homo est homo', et non facit mentionem de hoc quod dicunt, quod 'secundum quod' [ut] dicit habitudinem cuiuscumque cause indifferenter.

Ideo aliter est dicendum: verum est quod 'per se' in quarto modo differt a primo ut universale a particulare, non quia 'per se' in quarto modo dicat habitudinem ad quamlibet causam, sed quia omnes propositiones que sunt per se primo modo sunt per se quarto modo, quia propositiones

---

[a] Aristoteles, *Met.*, V, 18, 1022 a 14–22   [b] Aristoteles, *Met.*, V, 18, 1022 a 32–34

19 cause] a *add. sed del.* M   32 est modus] *iter.* M   45 modo] *iter.* M ‖ aliqua] alia M

3 It should be replied to this question that the fourth mode is a mode of inhering *per se*, for it counts as a mode of inhering *per se* whenever the subject is related to the predicate as cause to caused. In the fourth mode the subject is related to the predicate as cause to caused, and this is under the account of efficient cause. For instance, a killer kills because of killing, and there is no other cause of this, so it is obvious that the fourth mode is a mode of inhering. Hence the subject is related to its attribute in two genera of cause, namely the genus of material cause and the genus of efficient cause. As it is related to its attribute in the genus of material cause it is the second mode. As the subject is related to its attribute in the genus of efficient cause it is the fourth mode. Hence the subject as it is related under the account of efficient cause is a principle for demonstrating the attribute of its subject under the genus of material cause, for the subject is the effective cause of the attribute through its formal principles. Therefore the definition which joins the formal principles to the subject is the middle term for demonstrating the attribute of the subject, as will become apparent later. As for the subject at hand, both modes are modes of inhering, but the first mode of inhering falls among the premises of the demonstration, whereas the second falls into the conclusion.

But how does this mode differ from the other modes of inhering? Some say this fourth mode differs from the first and second as universal from particular. And they support this with Aristotle, where he gives the modes of being in a certain respect—the first mode is as it pertains to the form, the second as it pertains to matter. And then he adds that inasmuch as it indicates a relation (*habitudo*) to any cause, it is the fourth mode. In the same way, they speak of the *per se*, including the fourth mode of speaking per se.

But those who say this give too little consideration to the question. It is true that 'in a certain respect' in the first mode pertains to the form, and in the second mode to matter, but what these say, that in the fourth mode 'in a certain respect' indicates a relation to any cause, is false. For Aristotle holds that that is *per se* in the fourth mode for which there is no cause—in the way in which 'The human being is a human being' is *per se*. And he does not mention what these say, that 'in a certain respect' indicates a relation to any cause indifferently.

Therefore, it should be replied otherwise. It is true that *per se* in the fourth mode differs from the first as a universal from particular, not because *per se* in the fourth mode indicates a relation to any cause, but rather because every proposition that is *per se* in the first mode is *per se* in the fourth

per se primo modo sunt immediate et harum non est aliqua causa sicut propositiones per se in quarto modo.

Ad rationes.

Ad 1.1 Ad primam, concedo maiorem; ipsa enim causatio per se non est inherentia per se, quod tamen est causatum ab alio, potest ei per se inherere. Sed causatum ad causam potest dupliciter considerari: aut in abstractione aut in concretione. Si in abstractione, sic nec causa per se predicatur de causato nec e converso, sicut causatio non est per se inherentia. Sed in concretione dico quod causatum de causa per se potest predicari, ut 'interfectum interiit': predicatur causatum de causa per se in concretione, et propter hoc bene modus causandi est modus inherendi quantum ad concretionem. Ad minorem, si intelligat quod quartus modus sit ita modus causandi quod non inherendi, falsum dicit; nec videtur ratio quod quartus modus sit magis modus causandi quam secundus, quia sicut effectus in quarto modo predicatur de causa, sic in secundo modo materiatum de materia in genere cause materialis. Et si dicas quod sic, quia in quarto modo subiectum est causa efficiens predicati in subiecto, ⟨sed⟩ subiectum est causa materialis quia efficiens est causa causalitatis materie [quia efficiens], ideo quartus modus est modus causandi, secundus modus non est modus causandi, illud non valet, quia finis est causa efficientis. Et similiter forma et finis coincidunt, et si finis est causa causarum, ergo si predicatum comparatur ad subiectum sicut forma, maxime dicetur modus causandi, et talis est primus modus, et hoc est contra eos, et ita causa eorum non est sufficiens.

Ad 1.2 Ad aliam rationem, concedo maiorem. Et cum dicis: 'si predicatum comparetur ad subiectum ut causa ad causatum sic est primus modus', concedo. Et ulterius, cum dicitur quod cum predicatum comparatur ad subiectum ut causatum ad causam est secundus modus, dico quod hoc non est verum precise, quia causatum potest comparari ad causam vel in genere cause materialis, et sic est secundus modus, vel in genere cause efficientis, et sic est quartus modus. |

M 105ra

---

52 aliqua] alia M   56 inherentia] inherenter M   64 falsum] simul M   72 et^1] s *add. sed del.* M || finis^2] vis M

mode, for propositions *per se* in the first mode are immediate, and of these there is no cause, just as with propositions *per se* in the fourth mode.

In response to the arguments.

Ad 1.1 To the first, I grant the major premise, since causation *per se* is not inherence *per se*, whereas what is caused by another can inhere in it *per se*. But the relation of the caused to the cause can be considered in two ways, either in abstraction or concretely. If in abstraction, the cause is not predicated *per se* of the caused, nor vice versa. Thus causation is not *per se* inherence. But considered concretely, I hold that the caused can be predicated of the cause *per se*. For instance, in 'The killer kills', the caused is predicated of the cause *per se* and concretely. And for this reason a mode of causing certainly is a mode of inhering, considered concretely. In response to the minor premise, if Grosseteste thinks that the fourth mode is a way of causing so as not to be a way of inhering, then what he says is false. Nor does there seem to be any reason why the fourth mode is more a mode of causing than the second, since just as the effect in the fourth mode is predicated of the cause, so in the second mode the material thing is predicated of the matter in the genus of material cause. Perhaps you will say otherwise, because in the fourth mode the subject is the efficient cause of the predicate in the subject, whereas the subject is the material cause because the efficient cause is the cause of the causality of the matter, because it is efficient, and thus the fourth mode is a mode of causing and the second mode is not a mode of causing. This is not right, however, since the end is also an efficient cause, and form and end coincide similarly. And if the end is the cause of causes, then if the predicate is related to the subject as its form, then it will most surely be called a mode of causing and such is the first mode. This goes against them, and so their account of cause is not sufficient.

Ad 1.2 In response to the other argument, I grant the major premise, and when you say, 'If the predicate is related to the subject as cause to caused in this way then it is the first mode', I grant it. And next, when it is said that 'when the predicate is related to the subject as caused to cause' it is the second mode, I reply that this is not true precisely, for the caused can be related to the cause either in the genus of material cause, and then it is the second mode, or in the genus of efficient cause, and then it is the fourth.

⟨Questio 31⟩

Consequenter queritur qualiter modi dicendi per se se habeant ad demonstrationem.

1.1 Et videtur alicui[a] quod quartus modus non ingrediatur demonstrationem, quia contingit aliquando demonstrare passionem de subiecto per definitionem passionis; sed ubi accipitur definitio passionis pro medio, quartus modus non ingreditur demonstrationem, quia in tali demonstratione in maiori passio predicatur de sua definitione, qui modus reducitur ad primum; in secunda propositione definitio passionis predicatur de subiecto, et subiectum respectu passionis videtur se habere in ratione cause materialis; item, nec in conclusione, quia ibi predicatur passio de suo subiecto, et ita est secundus modus.

1.2 Item, contingit aliquando demonstrare passionem de subiecto [sicut] per aliam passionem priorem, sicut quod triangulus habet tres per hoc quod est figura habens angulum extrinsecum etc.; sed in tali demonstratione primus modus non ingreditur demonstrationem, quia in prima propositione passio de passione predicatur; nec in secunda, quia ibi prior passio predicatur de subiecto; nec in conclusione, quia ibi predicatur passio proposita de subiecto; et nullus istorum est primus modus.

3 Dicendum est ad hoc quod, cum quattuor sint modi eius quod est per se, quartus modus non ingreditur demonstrationem nisi sicut incomplexum aliquid quod per se determinat inherentiam. Verum est tamen quod, cum omne per se inherens dependeat ad aliquod per se ens, sicut omnis modus inherendi per se dependet ad modum essendi per se, [et] propter hoc in conclusione demonstrationis per se potissime, sicut ibi est modus inherendi per se, oportet quod subiectum sit ens per se. Et antequam deveniatur ad tale subiectum quod est ens per se de quo demonstratur passio de subiecto, non est demonstratio potissima. Unde non est universaliter verum dicere quod quartus modus non conferat ad demonstrationem, sed quod non conferat ut accipitur in premissis demonstrationis, hoc est verum de aliis modis.

---

[a] Cf. Egidius Romanus, *Super librum Posteriorum*, ad II, 9 (1488, f. 117D); cf. Albertus Magnus, *In Analytica Posteriora*, II, II, 10–11 (ed. Borgnet, pp. 188–93)

---

5 definitio passionis] passio demonstrationis M

Question 31.

Next it is asked in how the modes of speaking *per se* are related to demonstration.

1.1 It seems to some that the fourth mode does not enter into a demonstration, for it is possible sometimes to demonstrate an attribute of the subject through the definition of the attribute. But where the definition of an attribute is taken for the middle term, the fourth mode does not enter into the demonstration, for in the case of such a demonstration, (i) the attribute of the major premise is predicated of its definition, and this mode is reduced to the first mode of speaking *per se*; (ii) in the second proposition the definition of the attribute is predicated of the subject, and the subject with respect to its attribute seems to have the character (*ratio*) of the material cause; and again (iii) nor is the conclusion *per se* in the fourth mode, for there the attribute is predicated of its subject, and so this is *per se* in the second mode.

1.2 Furthermore, it is sometimes possible to demonstrate the attribute of the subject through another prior attribute, as with the fact that a triangle has three angles equal to two right angles through the fact that it is a figure which has an exterior angle etc. But in such a demonstration the first mode does not enter into the demonstration, for in the first proposition the attribute is predicated of an attribute, and in the second the prior attribute is predicated of the subject, and in the conclusion the attribute under consideration is predicated of the subject, and none of these is the first mode.

3 It should be replied to this that since there are four modes of being *per se*, the fourth mode does not enter into a demonstration except as some term that *per se* determines inherence. But it is true that, since every *per se* inherent depends on some *per se* being, so every mode of inhering *per se* depends on a mode of being *per se*. And because of this, in the conclusion of a demonstration *per se* of the strongest sort, just as there is a mode of inhering *per se* it is necessary that the subject be a being *per se*. And before such a subject is arrived at, which is a being *per se* of which the attribute is demonstrated, there is no demonstration of the strongest sort of the subject. Hence it is not always true to say that the fourth mode does not relate to demonstration, but that it does not relate as it is found in the premises of the demonstration. And this is true of the other modes.

Intelligendum est duo: primum est quod in demonstratione potissima passio demonstratur de subiecto; secundum est quod medium ad demonstrandum passionem de subiecto est definitio subiecti et dicens causam inherentie passionis ad subiectum. Et hoc est quod dicit Aristoteles,[a] secundo huius, quod medium dicit quid est et propter quid, dicit autem quid est subiecti et propter quid inherentie passionis ad subiectum. Unde in demonstratione potissima in maiori passio attribuitur medio, quod medium est causa efficiens passionis, quia medium in tali demonstratione dicit quid est subiecti et propter quid inherentie predicati ad subiectum, et sic est per se quarto modo. In secunda propositione definitio subiecti attribuitur subiecto, et ita est ibi primus modus. In conclusione passio attribuitur subiecto, et ita est ibi secundus modus. Et ita in demonstratione potissima quartus modus est in maiori, primus modus est in minore, secundus in conclusione, et sic in demonstratione potissima omnes modi requiruntur.

Ad argumenta.

Ad 1.1 Ad primum, cum arguitur 'passio aliquando ostenditur de subiecto per definitionem passionis, et ibi non est quartus modus dicendi per se', potest dici quod ibi ingreditur demonstrationem quartus modus, quia sicut subiectum se habet in ratione cause materialis respectu definitionis passionis, sic se habet in ratione cause efficientis, quia definitio passionis dicit essentiam passionis, essentia autem passionis causatur a subiecto effective, et ideo in tali demonstratione quartus modus sumitur in secunda propositione, quia in prima passio attribuitur definitioni passionis, in secunda autem definitio passionis subiecto.

Vel potest dici quod omnes tres modi ingrediuntur demonstrationem ubi est demonstratio potissima, ubi autem ostenditur passio de subiecto per definitionem passionis non est demonstratio potissima, quia talis

[a] Aristoteles, *AnPo*, II, 2, 90 a 7–23

36 inherentie passionis] *inv.* M 38 passionis] quia medium in tali demonstratione dicit quid est causa efficiens passionis quia medium in tali demonstratione dicit quid est subiecti et propter quid inherentie predicati ad subiectum et illud medium est causa efficiens passionis *add.* M 39 et] *iter.* M ‖ subiectum] et illud medium est causa efficientis passionis *add.* M 54 attribuitur] subiecto effective et ideo in tali demonstratione quartus modus *add. sed vac.* M

There are two things to be understood here. The first is that in a demonstration of the strongest sort an attribute is demonstrated of the subject. The second is that the middle term for demonstrating the attribute of the subject is the definition of the subject, which indicates the cause of the attribute's inherence in the subject. And this is what Aristotle says in the second book of the *Posterior Analytics*, that the middle term indicates what-it-is and because-of-what-it-is. But he is speaking of the what-it-is of the subject and the because-of-what of the attribute's inherence in the subject. Hence, in a demonstration of the strongest sort the attribute is attributed to the middle term in the major premise, which middle term is the efficient cause of the attribute, since the middle term in such a demonstration indicates the what-it-is of the subject and the because-of-what of the predicate's inherence in the subject. And so it is *per se* in the fourth mode. In the second proposition the definition of the subject is attributed to the subject, and so there is the first mode there. In the conclusion the attribute is attributed to the subject, and so there is the second mode there. And so in demonstration of the strongest sort the fourth mode is in the major premise, the first mode is in the minor premise, and the second is in the conclusion. Accordingly, the strongest sort of demonstration requires every mode.

In response to the arguments.

Ad 1.1 To the first, when it is argued that the attribute is sometimes shown of the subject through the definition of the attribute, and the fourth mode of speaking *per se* does not occur there, it can be held that the fourth mode enters into the demonstration there, since just as the subject has the character of a material cause with respect to the definition of the attribute, so it has the character of the efficient cause, since the definition of the attribute indicates the essence of the attribute, whereas the essence of the attribute is caused by the subject as efficient cause. Therefore in such a demonstration the fourth mode is assumed in the second proposition, since in the first the attribute is attributed to the definition of the attribute, but in the second the definition of the attribute is attributed to the subject.

Or it can be held that all three modes enter into a demonstration that is the strongest sort of demonstration, but where an attribute is shown of the subject through the attribute's definition it is not the strongest sort

demonstratio non procedit per immediata, quia adhuc potest ostendi definitio passionis de subiecto per aliquid prius.

Ad 1.2 Ad secundum, dicendum quod ibi primus modus non videtur ingredi demonstrationem, nihilominus si dicatur quod per primum intelligitur quartus, quia omnis propositio primo modo est per se quarto modo, ideo cum ibi sit quartus modus erit ibi primus ad minus sicut particulare in suo universali.

⟨Questio 32⟩

Consequenter queritur utrum aliqua negativa sit vera per se.

1.1 Et arguitur quod non, quia nulla propositio que solum est vera per accidens dicitur esse per se; sed omnis propositio negativa est vera per accidens, sicut dicit Aristoteles[a] in libro *Perihermeneias:* bonum non est malum nisi quia ista est vera, 'bonum est bonum'; cum ergo omnis negativa verificetur per affirmativam, videtur quod nulla negativa sit per se.

1.2 Item, hec est negativa vera 'nullus homo est asinus' et tamen non est per se, quia si esset per se, esset primo modo aut secundo aut quarto; sed non primo modo, quia predicatum non cadit in definitione subiecti; nec secundo modo, quia predicatum non est propria passio subiecti; nec quarto, quia subiectum non est causa efficiens predicati.

2 Oppositum arguitur: demonstratio est ex hiis que sunt per se; sed contingit facere demonstrationem ex aliquibus negativis; ergo etc.

3 Ad hoc est intelligendum quod in genere negativarum est devenire ad aliquam negativam per se et immediatam, quia si aliqua sit negativa | et debeat demonstrari, debet sumi aliqua negativa (et hoc non sufficit, sed debet sumi alia affirmativa); quero tunc de illa negativa probante, aut est immediata aut mediata, si immediata propositum habetur, si non, sed demonstretur per aliam, quero de illa alia aut erit immediata aut non, et sic in infinitum.

M 105rb

[a] Aristoteles, *Int.*, 14, 23 b 8–26

62 primum] quartum M   63 quartus] primus M   15 sit] s. l. M

of demonstration, since such a demonstration does not proceed through immediate propositions, since the attribute's definition can be shown of the subject through something prior.

Ad 1.2 As to the second, it should be replied that the first mode does not seem to enter into the demonstration there. Nevertheless, if it is said that the fourth is understood through the first, since every proposition in the first mode is *per se* in the fourth mode, then, since the fourth mode is there the first will be too, at least as a particular within its universal.

Question 32.

Next it is asked whether any negative proposition is true *per se*.

1.1 It is argued that none is, for no proposition that is true only accidentally is said to be *per se*. But every negative proposition is true accidentally, as Aristotle says in *De interpretatione*, 'the good is not bad only because this is true, "the good is good"'. Since, then, every negative proposition is made true by an affirmative proposition, it seems that no negative proposition is *per se*.

1.2 Furthermore, this is a true negative proposition, 'No human being is a donkey', and yet it is not *per se*, since if it were *per se* it would be so in the first mode, the second mode, or the fourth. But not in the first mode, since the predicate does not fall within the definition of the subject, nor in the second, since the predicate is not a proper attribute of the subject, nor in the fourth, since the subject is not an efficient cause of the predicate.

2 On the other hand, it is argued that demonstration is from what is *per se*, but it is possible to make a demonstration from negative propositions, therefore etc.

3 In reply to this it should be understood that some *per se* and immediate negative propositions are to be found in the genus of negative propositions, for if any proposition is negative and must be demonstrated, some negative proposition must be assumed (and this does not suffice by itself, but some other affirmative proposition must also be assumed). I ask, then, about that negative proposition from which it is proved whether it is immediate or not. If it is immediate then we have what was proposed, and if not then it is demonstrated from some other negative proposition, and I ask about it whether it will be immediate or not, and so on indefinitely.

Item, si negativa non possit probari nisi sumatur affirmativa, sed in affirmativis est status, ergo in negativis; sed non esset ibi status nisi aliqua esset immediata; ergo etc. Minor probatur per Aristotelem[a] in hoc primo, scilicet quod sit status in affirmativis tam in predicatis essentialibus quam in accidentalibus.

Item, in unoquoque genere quod est per accidens reducitur ad illud quod est per se in illo genere; sed in genere negativarum contingit reperire aliquam negativam per accidens; ergo ibi erit aliqua per se ad quam reducitur.

Item, sicut in affirmativis est reperire aliquam primam, sicut in terminis entis, sic in genere negativarum contingit reperire aliquam primam qua non est alia notior, et ista fit in terminis communibus entis, puta 'non ens non est ens' vel 'ens non est non ens'.

Sed tu dices: si sit dare negativam immediatam, cum immediatum sit quo non est aliud prius et notius, erit tunc aliqua negativa qua non erit aliqua prior, et hoc est contra Philosophum,[b] qui dicit quod omni negativa est affirmativa prior. Dicendum quod sicut aliquid est immediatum, sic nihil est prius eo; sed aliquid esse immediatum est dupliciter: vel simpliciter vel in genere. Unde si aliquid sit immediatum in genere, nihil est prius in genere illo. Unde contingit devenire ad aliquam negativam primam in genere negativarum, et tali immediato est aliquid prius.

Sed tu dices: cum nulla negativa sit immediata simpliciter, quoniam est aliquid prius, videtur quod demonstrationes non fiant ex negativis, cum demonstratio fiat ex immediatis simpliciter. Concedo quod nulla demonstratio potissima fit ex negativis. Unde notandum est quod non omnis demonstratio fit ex affirmativis primis simpliciter, sed alia fit ex affirmativis immediatis in suo genere; si enim demonstratio semper fieret ex immediatis simpliciter, non esset nisi una demonstratio, et sic de negativis.

Sed tu queres ulterius, si alique negative sint per se, quomodo reperientur in eis modi dicendi per se? Intelligendum est quod, sicut in affirmativis

---

[a] Aristoteles, *AnPo*, I, 7, 75 b 2; cf. I, 22, 83 b 20   [b] Aristoteles, *Int.*, 14, 23 b 8–26

Further, if a proposition is negative it cannot be proved unless an affirmative proposition is assumed. But there is a stopping place in affirmative propositions, and so too in negative propositions. But there would be no stopping place there unless some negative proposition is immediate, therefore etc. The minor premise is proved from Aristotle in *Posterior Analytics* I, namely that there is a stopping place in affirmatives, and in essential predicates as well as accidental predicates.

Further, in each genus what is accidental is reduced to what is *per se* in that genus, but in the genus of negative propositions it is possible to find an accidental negative. Therefore there will be something *per se* in that genus to which it is reduced.

Further, just as among affirmative propositions there is some first to be found, for instance, one put in terms of being, so in the genus of negatives it is possible to find a first in comparison to which no other is better known, and this will be found in the common terms of being, for instance, 'a non-being is not a being', or 'a being is not a non-being'.

But you will say, if there is an immediate negative proposition, since an immediate proposition is one to which no other is prior and better known, there will then be some negative to which nothing is prior, and this is contrary to the Philosopher, who says that to every negative proposition some affirmative proposition is prior. It should be replied that if anything is immediate then nothing is prior to it, but 'something is immediate' is said in two ways: either without qualification or in a genus. Hence if something is immediate in a genus, nothing is prior to it in that genus, and so it is possible to arrive at a first negative in the genus of negative propositions and yet for something to be prior to such an immediate proposition.

But you will say, since no negative is immediate without qualification, because there is something prior, it seems that demonstrations do not arise from negatives, since a demonstration arises from propositions immediate without qualification. I grant that no demonstration of the strongest sort is from negative propositions. It should be noted, then, that not every demonstration is from affirmative first principles without qualification. There are others from affirmative propositions immediate in their genus. For if a demonstration always arose from immediate propositions without qualification, then there would be no more than one demonstration, and so there would be no demonstration in the case of negatives.

But you will ask next whether any negatives are *per se* in any mode that is to be found among these modes of speaking *per se*. It should be understood

est assertio predicati ad subiectum, sic in negativa est deassertio predicati ad subiectum. In aliquibus negativis tale est subiectum quod ipsum in sua definitione includit oppositum predicati, ut hic 'nullus homo est asinus', et tales negative sunt per se primo modo. Aliquando autem subiectum est tale quod ab eo removetur predicatum, ita quod subiectum est proprium subiectum respectu oppositi predicati, ut hic 'homo non est irrisibilis', et sic est secundus modus. Aliquando autem contingit quod subiectum sit tale quod sit causa efficiens oppositi predicati, ut hic 'nullum interfectum vivit', et sic est quartus modus.

Ad rationes.

Ad 1.1 Ad primam, cum arguitur 'nulla propositio etc.', dico quod aliquid esse per accidens est dupliciter: vel per accidens quod opponitur ei quod est per se vel ei quod est primo. Unde concedo quod nulla propositio est ita per accidens quod opponitur ei quod est per se [est per se], propositio tamen per accidens quod opponitur ei quod est primo potest esse per se; multe enim sunt propositiones per se que non sunt primo, tales enim sunt negative.

Ad 1.2 Ad aliam rationem, dico quod ista 'nullus homo est asinus' est per se primo modo, quia predicatum removetur a subiecto, ita quod oppositum predicati includitur in ratione subiecti.

⟨Questio 33⟩

Consequenter queritur de 'per se' in comparatione ad 'secundum quod ipsum', primo utrum 'per se' et 'secundum quod ipsum' sint idem.

1.1 Et arguitur quod sic per Philosophum,[a] qui hoc dicit; dicit enim quod 'per se' et 'secundum quod ipsum' idem dico esse; triangulus enim secundum quod ipsum habet tres angulos equales duobus rectis ⟨et trianguli per se anguli duobus rectis⟩ equales sunt; quare etc.

---

[a] Aristoteles, *AnPo*, I, 4, 73 b 29

that, as in affirmative propositions there is an assertion of the predicate to the subject, so in a negative proposition there is a denial of a predicate to the subject. In some negative propositions the subject is such that it includes the opposite of the predicate in its definition, as for instance in this proposition, 'No human being is a donkey'. And such negative propositions are *per se* in the first mode. Sometimes the subject is such that the predicate is removed from it in such a way that the subject is a proper subject with respect to the opposite predicate, for instance in this, 'man is not unable to laugh'. So it is the second mode. And sometimes it is possible that the subject is such that it is the efficient cause of the opposite of the predicate, as for instance, in 'nothing killed lives'. And this is the fourth mode.

In response to the arguments.

Ad 1.1 To the first, when it is argued that 'no proposition' etc., I hold that there are two ways for something to be accidentally, either that accidentally which is opposed to what is *per se*, or that which is opposed to what is primary. Hence I grant that no proposition is accidental in the way that is opposed to that which is *per se*, but a proposition that is accidental by being opposed to what is primary can be *per se*. For there are many *per se* propositions that are not primary, for negative propositions are like that.

Ad 1.2 In response to the other argument, I reply that this proposition, 'no human being is a donkey', is *per se* in the first mode, for the predicate is removed from the subject in such a way that the opposite of the predicate is included in the account of the subject.

Question 33.

Next it is asked about *per se* in comparison to 'according to itself' (*secundum quod ipsum*). First, it is asked whether *per se* and 'according to itself' are the same.

1.1 It is argued that they are the same according to the Philosopher, who says this. For he says 'I say that "*per se*" and "according to itself" are the same thing'. For a triangle according to itself has three angles equal to two right angles, and the angles of a triangle are equal *per se* to two right angles. Therefore etc.

1.2 Item, illa que significant idem sunt eadem; sed 'per se' et 'secundum quod ipsum' significant idem; ergo etc. Maior patet. Minor probatur, quia idem significatur per istam propositionem 'homo per se est animal' et 'homo secundum quod ipsum est animal'.

2.1 Oppositum arguitur sic: secundum quod ipsum inesse et primo inesse sunt idem; ergo etc.; non enim omne quod per se inest primo inest.

2.2 Item, si 'per se' et 'secundum quod ipsum' sunt idem, ubicumque invenitur unum, et aliud; sed hoc est falsum, quia dicendo sic 'ysocheles habet tres' est per se, non tamen secundum quod ipsum, quia ysocheles secundum quod ysocheles non habet tres, sed secundum quod triangulus.

3 Dico quod 'per se' et 'secundum quod ipsum' aliquo modo sunt idem, aliquo modo sunt diversa. Sunt idem quia 'per' dicit causam et 'se' est pronomen reflexivum, et ideo 'per se' duo includit, scilicet relatum identitatis et nomen causalitatis, et ideo ubicumque invenitur in aliqua propositione, denotetur esse causalitatis, et illa causalitas est subiecti supra predicatum, quia 'se' est pronomen reflexivum, ideo solum refert illud quod precedit, et ideo solum refert subiectum et non refert quamcumque causalitatem sed solum illam. Similiter, 'secundum quod ipsum' duo includit, scilicet relatum identitatis et nomen causalitatis. In hoc ergo conveniunt ista duo, scilicet 'per se' et 'secundum quod ipsum'. Quod utrumque importat causalitatem patet per Philosophum[a] V *Metaphysice*; ibi enim ponit tres modos 'secundum quod' magis usitatos. Primus modus est prout essentia vel forma rei est illud per quod aliquid est, sicut | idea boni est illud per quod aliquid est bonum. Secundus modus est prout 'secundum quod' dicitur subiectum in quo natum est fieri aliquid primo, et secundum hoc dicimus quod corpus est coloratum secundum superficiem, quia superficies est illud secundum quod corpus est coloratum primo. Tertius modus dicitur prout 'secundum quod' denotat causam, id est cuius causa venit aliquis, et iste tertius modus reducitur ad quartum hic. Unde in quarto modo non solum predicatum comparatur ad subiectum in ratione cause efficientis sed etiam finalis. Sic ergo inducendo patet quod 'secundum quod' importat causalitatem et qualiter sunt idem.

M 105va

---

[a] Aristoteles, *Met.*, V, 18, 1022 a 14–24

---

9 istam] ipsam M 10 quod ipsum] *post. corr.* M 19 reflexivum] relativum M 22 quia] quod M || se] *s. l.* M || reflexivum] relativum M 34 causam] aliquis M 38 et] secundum quod ipsum *add. sed del.* M

BOOK I

1.2 Furthermore, those that signify the same are the same, but *'per se'* and 'according to itself' signify the same, therefore etc. The major premise is obvious, and the minor premise is proved because the same is signified through the propositions 'a human being is *per se* an animal', and 'a human being according to itself is an animal'.

2.1 On the other hand, it is argued that to inhere in something according to itself and to inhere in it primarily are the same, therefore etc. For not everything that is in something *per se* is in it primarily.

2.2 Furthermore, if *per se* and 'according to itself' are the same, wherever they are found one is the other. But this is false, since the claim that 'an isosceles triangle has three angles equal to two right angles' is *per se*, but not according to itself, since an isosceles triangle does not have three angles etc. considered as an isosceles triangle, but only considered as a triangle.

3 I reply that *per se* and 'according to itself' are in one way the same and in another way different. They are the same because *'per'* indicates cause, and *'se'* is a reflexive pronoun, and so *per se* includes two things, namely the relation of identity and a term for causality, and so wherever it is found in any proposition, the being of causality is denoted and that the causality is that of the subject upon the predicate. This is because *'se'* is a reflexive pronoun. Therefore it only refers to that which precedes, and so it refers only to the subject, and does not refer to any other causality, but only to that one. In a similar way 'according to itself' includes two things, namely the relation of identity and a term for causality. In this, therefore, these two, namely *per se* and according to itself, agree. That both convey causality is obvious from the Philosopher in *Metaphysics* V, where he posits three ways in which 'according to' is most often used. The first way is as the essence or form of a thing is that through which something is, as the idea of good is that through which something is good. The second way is as 'according to' indicates the subject in which something is suited by nature to arise primarily, and in this way we say that body is colored considered as surface since the surface is that according to which body is colored primarily. The third way is said as 'according to' denotes a cause, that is, the cause why someone comes, and this third way is reduced to the fourth here. Whence in the fourth way not only is the predicate related to the subject in the account of efficient cause, but also of final cause. Therefore, by working through the cases in this way, it is clear that 'according to' conveys causality and how they are the same.

Item, aliquo modo sunt diversa. Ad cuius evidentiam considerandum est quod differenti modo sunt ista per se: 'triangulus habet tres' et 'ysocheles habet tres', quia triangulus secundum totum illud quod est habet tres, per sua principia est causa huius quod est habere tres, unde secundum quod triangulus habet tres; sed licet ysocheles secundum totum id quod est habeat tres, non tamen habere tres inest sibi primo, sed prius triangulo. Ista ergo est per se, non tamen secundum quod ipsum, quia si secundum quod ysocheles habere⟨t⟩ tres, primo inesse⟨t⟩ sibi et per ipsum inesse⟨t⟩ omni alii; hoc autem est falsum, quia triangulus habet tres per se et primo, non tamen per naturam ysochelis sed per naturam trianguli.

Sed aliquis diceret quod videtur accipi falsum in ratione, quia dicitur quod si ysocheles in quantum ysocheles haberet tres per se et primo, ⟨habere tres⟩ inesset omni alii per naturam ipsius. Dicimus enim quod ista est per se et primo 'homo est animal', tamen non dicimus quod animal inest omni alii per hominem; ergo etc.

Aliquis diceret quod locutio est impropria, 'homo secundum quod ipsum est animal', ista tamen est propria, 'homo secundum se est animal'; ista enim differunt, 'secundum se' et 'secundum quod ipsum'. Tunc enim dicitur aliquid esse aliquid secundum se quando illud quod inest et illud cui inest non differunt realiter; ideo ista [non] est propria: 'homo est animal secundum se'. Tunc dicitur aliquid ⟨esse aliquid secundum quod ipsum⟩ quando illud quod inheret et illud cui inheret realiter differunt, ita quod unum sit accidens immediatum causatum ab essentia alterius; et ita non est propria 'homo secundum se est risibilis', sed ista: 'homo secundum quod ipsum ⟨est risibilis⟩'.

Adhuc diceret aliquis: dicitur quod ista est per se 'ysocheles habet tres', aut ergo primo modo, quod non potest, quia predicatum non cadit in definitione subiecti, nec secundo modo [nec secundo modo], quia in propositione per se secundo modo predicatum solum convenit subiecto, vel si aliis inest, eis inest per naturam ipsius subiecti; tale autem non est ysocheles respectu huius quod est habere tres.

Dico quod in primo modo et secundo sunt diversi gradus, quia in primo per se aliqua est per se et primo, aliqua per se non primo, sicut ista est

---

43 totum] hi *add. sed del.* M   55 animal¹] propria M   57 aliquid²] aliud M   58 propria] secundum se M   59 se] quod ipsum M   62 et] *iter. sed corr.* M   63 ipsum] adh *add. sed del.* M   71 aliqua¹] alia M

Furthermore, in a way they are different. To make this evident one should consider that these are *per se* in different ways, 'the triangle has three etc.' and the isosceles has three etc.'. For a triangle according to all that it is has three angles equal to two right angles, through its principles it is the cause of its having three etc., and hence considered as a triangle it has three etc. But although an isosceles, according to all that it is, has three etc., having three etc. does not inhere in it primarily, but is prior in triangle. So this is *per se*, but not according to itself, since if it were true according to its being isosceles, then having three etc. would inhere primarily in it and through this it would inhere in every other case. But this is false, since a triangle has three etc. *per se* and primarily, and not through the nature of isosceles, but rather through the nature of triangle.

But someone might say that the argument seems to presuppose something false, since it is said that if an isosceles inasmuch as it is isosceles should have three etc. *per se* and primarily, then having three etc. would inhere in every other case through its nature. For we say that this holds *per se* and primarily, 'Human being is animal', but we do not say that animal inheres in every other case through human being; therefore etc.

Someone might say that this locution is improper, 'Human being according to itself is animal', but this is proper, 'Human being in its own right (*secundum se*) is animal', for these two, in its own right (*secundum se*) and according to itself (*secundum quod ipsum*), differ. For then it is said that something is something in its own right when that which inheres and that in which it inheres do not differ in reality. Therefore this locution is proper: 'Human being according to itself is animal'. Then it is said that something is something according to itself when that which inheres and that in which it inheres differ in reality, so that one is an immediate accident caused by the essence of the other, and so this is not proper, 'Human being in its own right is able to laugh', but this is, 'Human being according to itself is able to laugh'.

Moreover, someone might say, it is said that this is *per se*, 'an isosceles has three etc.', so that it must be either in the first mode, which cannot be since the predicate does not fall in the definition of the subject, nor can it be in the second mode, since in a proposition *per se* in the second mode the predicate only agrees with the subject, or if it inheres in others it inheres in them through the subject's nature, but isosceles is not such a subject with respect to having three etc.

I reply that in the first and the second there are different degrees, since in the first mode of speaking *per se* some propositions are *per se* and primary, and some *per se* and not primary, as this is *per se* and primary, 'A human

per se et primo: 'homo est animal', sed ista est per se [est per se] non primo: 'Sortes est animal'. Eodem modo ⟨in secundo per se⟩ aliqua est per se et primo, ut 'triangulus habet tres', ⟨aliqua est per se non primo, ut 'ysocheles habet tres'⟩. Sed quomodo inerit aliquid alicui per se secundo modo, nisi sit causatum ex principiis propriis et essentialibus ipsius subiecti? Ideo, [quia] hec est per se primo: 'triangulus habet tres', quia habere tres causatur ex principiis essentialibus trianguli, sed ad hoc quod sit per se secundo ⟨modo⟩ non oportet quod per se causetur ex propriis et essentialibus principiis subiecti, sed sufficit quod subiectum participet essentialiter naturam illius quod est causa passionis per sua principia essentialia et propria cui per se et primo inest, ut ysocheli inest habere tres et secundo, et ideo non oportet quod habere tres causetur immediate a propriis principiis ipsius, sed sufficit quod ipsum essentialiter pertinet ⟨ad⟩ naturam ipsius cui primo et per se ipsum inest. Et sic patet qualiter 'per se' et 'secundum quod ipsum' sunt aliquo modo eadem, aliquo modo diversa.

Sed tu queres, ubi ponemus ipsa esse eadem et ubi diversa? Dico quod in illis que sunt per se et primo sunt eadem 'per se' et 'secundum quod ipsum', sed in illis que sunt per se non primo 'per se' et 'secundum quod ipsum' sunt diversa.

Rationes procedunt viis suis.

⟨Questio 34⟩

Consequenter queritur utrum aliquod accidens quod primo et universaliter inest alicui subiecto possit demonstrari de eo.

1.1 Et arguitur quod non, quia illius quod demonstratur est altera causa, si ergo passio que primo inest subiecto demonstratur de subiecto demonstrabitur per causam alteram; ergo illa passio prius inerit illi cause quam subiecto, et hoc est contra positum, quia ponitur quod primo inest subiecto; ergo etc.

1.2 Item, quod immediate inest alicui non potest demonstrari de eo, quia immediatorum non est demonstratio; sed passio que primo et per se inest subiecto inest ei immediate; ergo etc.

---

73 aliqua] alia M

being is an animal', but this is *per se* and not primary, 'Socrates is an animal'. In the same way, some propositions in the second mode are *per se* and primary, as 'A triangle has three' etc., whereas some are *per se* but not primary, as 'an isosceles triangle has three' etc. But how can something inhere in something *per se* in the second mode, unless it is caused from the proper and essential principles of the subject? Therefore, this is *per se* and primary, 'A triangle has three etc.', because its having three etc. is caused by the essential principles of triangle. But in the case of the second mode of speaking *per se*, it is not necessary that it be caused *per se* by the proper and essential principles of the subject, rather it suffices that the subject participate essentially in the nature of that which is the cause of the attribute through its essential and proper principles, which inhere in it *per se* and primarily. For instance, having three etc. inheres in isosceles in the second mode, and therefore it is not necessary that having three etc. be caused immediately by its proper principles, rather it suffices that it pertain essentially to the nature of that which it inheres in primarily and *per se*. And thus it is obvious how *per se* and according to itself are in a way the same, and in a way different.

But you will ask where we posit these to be the same, and where different. I reply that in those which are *per se* and primary, *per se* and 'according to itself' are the same, but in those which are *per se* and not primary, *per se* and 'according to itself' are different.

The arguments each go through in their own way.

Question 34.

Next it is asked whether any accident which inheres in a subject primarily and universally can be demonstrated of it.

1.1 It is argued that it cannot, for of that which is demonstrated there is a further cause. If, therefore, an attribute that inheres in a subject primarily is demonstrated of the subject, it will be demonstrated through some further cause. Therefore that attribute's inherence in that cause will be prior to its inherence in the subject, and this is contrary to what was assumed, for it was assumed to inhere in the subject primarily. Therefore etc.

1.2 Furthermore, what inheres immediately in something cannot be demonstrated of it, for of immediate things there is no demonstration. But an attribute that is in a subject primarily and *per se* is in it immediately. Therefore etc.

1.3 Item, si passio que primo inest alicui subiecto demonstretur de eo, aut hoc erit per aliam passionem priorem, aut per passionem posteriorem, aut per definitionem passionis, aut per definitionem subiecti. Non potest demonstrari per passionem priorem, quia positum est quod illa sit prima; nec per posteriorem, quia tunc esset petitio principii; nec per definitionem passionis, quia illa est demonstrabilis de subiecto sicut passio; nec per definitionem subiecti, quia oportet medium esse eiusdem coordinationis cum extremis, sed si definitio subiecti acciperetur pro medio, medium non esset eiusdem coordinationis cum extremis, quia medium | esset de genere substantie et extrema de genere accidentium; et non potest pluribus modis demonstrari passio de subiecto; quare etc.

M 105vb

2 Oppositum patet per Philosophum[a] in littera; dicit enim quod tunc universale est cum demonstratum est aliquid quod inest subiecto per se et primo. Vult ergo quod passio que inest ⟨per se⟩ et primo subiecto possit demonstrari de subiecto.

3 Ad hoc est dicendum quod passio que per se et primo inest subiecto potest demonstrari de subiecto, et eius proprie est demonstratio, et alia non demonstrantur nisi ⟨per⟩ illam, quia passio que primo inest subiecto causatur ex principiis essentialibus subiecti, quia in compositis materia cum forma est causa cuiuslibet accidentis, cum hec passio non sit de essentia subiecti, habebit causam per quam inest subiecto. Tunc arguo: omne quod habet aliquam causam quare inest subiecto potest demonstrari de subiecto; sed omnis passio que inest subiecto per se et primo habet causam in subiecto, sicut universaliter omne accidens; ergo etc.

Item, definitio subiecti medium est in demonstratione potissima, sicut videbitur in secundo huius.[b] Cum ergo passio prima subiecti sit causata a subiecto per definitionem subiecti, ergo definitio subiecti poterit esse medium ad demonstrandum passionem de subiecto. Quod definitio subiecti sit causa illius passionis patet, quia definitio subiecti dicit quod quid est subiecti, et subiectum per suum quod quid est est causa passionis in ipso, quare ipsa definitio subiecti dicit causam passionis in subiecto.

[a] Aristoteles, *AnPo*, I, 4, 73 b 25–28 [b] Cf. Aristoteles, *AnPo*, II, 11, 94 a 24–36

25 subiecto] substantia M   32 aliquam] aliam M

1.3 Furthermore, if an attribute that is primarily in a subject is demonstrated of it, either this will be through another prior attribute, or through some posterior attribute, or through the definition of the attribute, or through the definition of the subject. It cannot be demonstrated through a prior attribute, since it is assumed that it is primary; nor through a posterior attribute, since that would beg the question, nor through a definition of the attribute, since that is demonstrable of the subject as an attribute of it; nor through a definition of the subject, since it is necessary that the middle term be in the same order as the extremes, and if the definition of the subject were taken as the middle term, then the middle term would not be in the same order with the extremes, for the middle term would be in the genus of substance, and the extremes in the genus of accident. But there cannot be any further ways to demonstrate the attribute of the subject. Therefore etc.

2 The contrary view is clear from the Philosopher in the text, for he says that something is universal when it is demonstrated that it inheres in the subject *per se* and primarily. Therefore he means that the attribute, which also inheres *per se* and primarily in the subject, can be demonstrated of the subject.

3 In response to this, it should be held that an attribute that is in a subject *per se* and primarily can be demonstrated of the subject, and this is a demonstration strictly speaking, and nothing else is demonstrated except through that ⟨kind of demonstration⟩. For an attribute which inheres primarily in its subject is caused by the essential principles of the subject, for in composites the matter together with the form is the cause of every accident, and since this attribute is not of the essence of the subject, it will have a cause through which it inheres in the subject. Then, I argue, everything that has some cause on account of which it is in the subject can be demonstrated of the subject, but every attribute that inheres in a subject *per se* and primarily has a cause in the subject, just as every accident does, universally. Therefore etc.

Furthermore, the definition of the subject is the middle term in a demonstration of the strongest sort, as we shall see in the second book. Since, then, a primary attribute of a subject is caused by the subject through the definition of the subject, therefore the definition of the subject can be the middle term for demonstrating the attribute of the subject. That the definition of the subject is the cause of the attribute is clear, since the definition of the subject indicates the essence (*quod quid est*) of the subject, and the subject through its essence is the cause of the attribute in itself. Therefore the definition of the subject indicates the cause of the attribute in the subject.

Et est intelligendum ulterius quod accidentia sint triplicia, scilicet prima, media et ultima. Accidentia prima possunt demonstrari de subiecto per causam existentem in subiecto, media autem possunt demonstrari per prima et ultima per media. Et ultima solum sunt demonstrabilia et non possunt esse media ad demonstrandum aliquid aliud ex quo ipsa sunt ultima, sicut apparet in naturalibus. Ignis enim per suam formam substantialem est causa caliditatis, et per caliditatem est causa levitatis et raritatis, et sic usque ad aliquam passionem que est demonstrabilis solum oportet devenire, et ipsa ulterius non est medium, ita quod in omnibus universaliter oportet querere statum, tam in naturalibus quam in aliis.

Ad rationes.

Ad 1.1 Ad primam, cum arguitur 'si accidens ostendatur de subiecto, ergo per aliud, [et illi prius inherebit]; ergo prius inerit illi', verum est sub ratione cause, sed non prius inerit illi sub ratione subiecti; unde habere tres inest triangulo per causam, et illi cause prius inest ut cause, non tamen ut subiecto, et ita potest passio demonstrari per causam.

Ad 1.2 Ad aliam rationem, dico quod aliquid esse immediatum alii est dupliciter, aut immediatum immediatione subiecti aut immediatum immediatione cause. Unde omnis propria passio immediate inest subiecto immediatione subiecti, non tamen immediatione cause; immo omnis passio habet causam ad demonstrationem sui de subiecto, licet primo et per se insit subiecto. Quod autem est immediatum alii immediatione cause [et immediatione] non potest demonstrari de subiecto.

Ad 1.3 Ad tertiam rationem, dicitur quod per definitionem subiecti ostenditur talis passio de subiecto. Et cum dicitur 'medium debet esse eiusdem coordinationis cum extremis', dico quod non oportet quod medium sit eiusdem coordinationis predicamentalis cum extremis, sed sufficit quod sit eiusdem coordinationis loquendo de tali coordinatione que est inter causam et causatum; talis autem est inter definitionem subiecti et ipsam passionem que primo et universaliter inest subiecto, que definitio subiecti est causa et passio ipsa causatum, et hec identitas coordinationis medii cum extremis sufficit.

---

45 et^1] sed M   46 aliquid] aliquod M   48 caliditatis – caliditatem] causalitatis – causalitatem M   61–62 passio] demonstra *sic* M

And it should be understood next that accidents are three-fold: primary, middle, and last. Primary accidents can be demonstrated of the subject through a cause that exists in the subject, whereas middle accidents can be demonstrated through primary accidents, and the last through middle accidents. Last accidents are only demonstrable, and cannot be middle terms for demonstrating anything else, which is why they are last. This is apparent in nature, for fire is the cause of heat through its substantial form, and through its heat it is the cause of lightness and rareness. In this way one must arrive at some attribute that is demonstrable only, and this last attribute is not a middle term, and thus in everything, universally, one has to seek a stopping point, both in natural cases and in others.

In response to the arguments.

Ad 1.1 In response to the first, when it is argued, 'if an accident is shown of its subject, then this is through another, and therefore it will be prior in belonging to it', this is true with respect to its being a cause, but it will not be prior in belonging to it but it will not be prior to it with respect to being a subject. So, having three angles etc. belongs to a triangle through a cause, and its belonging to that cause is prior as a cause, but not as its subject. And so an attribute can be demonstrated through its cause.

Ad 1.2 In response to the second argument, I reply that something's being immediate to another occurs in two ways: either immediate by its immediacy to its subject or immediate by the immediacy of its cause. Hence every proper attribute inheres immediately in its subject by the immediacy of the subject, but not by the immediacy of its cause. Indeed, every attribute has a cause of its demonstration in the subject, even though it inheres primarily and *per se* in the subject. But what is immediate in another by the immediacy of its cause cannot be demonstrated of the subject.

Ad 1.3 In response to the third argument, I hold that through the definition of the subject such an attribute is shown of the subject; and when it is said 'the middle term must be in the same order with the extremes', I reply that it is not necessary that the middle term be in the same order as the extremes as it relates to its category, but it suffices that it be of the same order speaking of such an order as exists between cause and caused. But such an order exists between the definition of the subject and the attribute that primarily and universally inheres in the subject, an order by which the definition of the subject is the cause and the attribute is the thing caused, and this identity in order of the middle term with the extremes is sufficient.

⟨Questio 35⟩

Consequenter queritur utrum hec sit universalis: 'homo est animal', universalitate determinata in hoc libro.[a]

1.1 Et arguitur quod sic, quia Themistius[b] super capitulum de erroribus dicit quod genus universaliter predicatur de suis speciebus; sed animal est genus ad hominem, quare universaliter inest homini; ideo ista ⟨est⟩ universalis universalitate determinata in hoc libro.

1.2 Item, illa propositio est universalis in qua predicatum inest subiecto universaliter per se et primo; sed in hac propositione 'homo est animal' animal inest homini universaliter, quia hec est vera: 'omnis homo est animal', et est per se, quia predicatum est de ratione subiecti. Item, ⟨in⟩est primo, quia illud predicatum inest primo quod non inest per aliquod prius; sed animal inest homini non per aliquod prius; ergo etc.

2 Oppositum arguitur: in propositione universali, ut hic loquimur de universali, predicatum non excedit subiectum nec e converso; sed animal excedit hominem, in pluribus enim reperitur quam homo; ergo etc.

3 Ad hoc intelligendum quod universale accipitur quattuor modis: uno modo secundum causalitatem, alio modo secundum predicationem, alio modo universale idem est quod propositio universalis, quarto modo dicitur universale convertentia predicati cum subiecto. Universale secundum causalitatem est illud cuius causalitas se extendit ad productionem plurium effectuum, et isto modo dicitur sol universale; est enim causa omnium generabilium et corruptibilium. Alio modo dicitur universale secundum predicationem, et illud est causatum per operationem intellectus | abstrahentis aliquid a conditionibus materialibus sive individuantibus, et de hoc dicit Aristoteles:[c] universale est unum in multis et de multis, et de tali non intendimus hic. Tertio modo universale quod est propositio universalis, ut ista 'omnis homo est animal', et adhuc isto modo non intendimus. Quarto modo dicitur universale ipsa convertentia predicati cum subiecto, et hoc modo dicimus quod universale inest universaliter sue proprie passioni, quia subiectum et sua passio non sunt sicut excedens et

M 106ra

---

[a] Aristoteles, *AnPo*, I, 4, 73 b 25–28   [b] Cf. Themistius, *Analyticorum posteriorium paraphrasis*, 4 (ed. O'Donnell, p. 257)   [c] Cf. Aristoteles, *Int.*, 7, 17 a 37–38

---

9 omnis] *iter.* M   12 sed animal] *iter.* M   21 effectuum] effectum non M   24 individuantibus] indivisibilibus M

Question 35.

Next it is asked whether 'a human being is an animal' is universal by the universality treated in this book.

1.1 It is argued that it is, for Themistius, on the chapter about errors, says that the genus is predicated universally of its species, but animal is the genus of a human being, so it universally inheres in human being. Therefore this is universal by the universality treated in this book.

1.2 Furthermore, that proposition is universal in which the predicate inheres in the subject universally, *per se*, and primarily. But in this proposition, 'a human being is an animal', animal inheres in human being universally, since this is true, 'Every human being is an animal'. And it is *per se*, for the predicate is of the account of the subject. Again, it inheres primarily, since that predicate inheres primarily which does not inhere through something prior; but animal does not inhere in human being through anything prior. Therefore etc.

2 On the other hand, it is argued that in a universal proposition, as we mean here by 'universal', what is predicated does not exceed the subject, nor vice versa; but animal exceeds human being, for it is found in many things other than human being; therefore etc.

3 In response to this, it should be understood that 'universal' can be taken in four ways, in one way according to causality, in a second way according to predication, in a third way a universal is the same as a universal proposition, and in the fourth way 'universal' is said of a proposition in which the subject and predicate are convertible. A universal according to causality is that by which causality extends to the production of many effects, and in this way the sun is called universal, for it is the cause of all generable and corruptible things. In the second way, 'universal' is said according to predication, and is caused through the operation of the intellect abstracting something from material or individuating conditions. And Aristotle says about this that a universal is one in many and from many. We are not considering such universals here. In the third way, a universal is a universal proposition, for instance, 'Every human being is an animal'. We are not considering universal in this way here. In the fourth way, a proposition in which the subject and predicate are convertible is universal, and in this way we say that the universal inheres in the

excessum, ⟨sed⟩ in quibus unum reperitur et reliquum et e converso. Et sic intendit Philosophus hic de universali. Unde dicit propositionem esse universalem in qua predicatum inest subiecto per naturam subiecti cum quo convertitur, ita quod non per aliquod aliud sibi inest.

Ad questionem igitur dicendum quod ista, 'homo est animal', in uno convenit cum universale, sed in tribus disconvenit. In hoc enim convenit cum ipso, quia animal inest homini universaliter, per se et primo, ut ostensum est. Sed in tribus disconvenit: primo quia in propositione in qua predicatum primo inest subiecto, ipsum non inest alii nisi per naturam ipsius prout hic sumitur 'primo', taliter autem non convenit animal homini; adeo enim equaliter convenit asino sicut et homini. In alio etiam disconvenit, quia in propositione universali predicatum est propria passio subiecti, et non sunt sicut excedens et excessum, sic autem non se habent animal et homo, quia animal non est propria passio hominis nec est convertibile cum homine; dicitur enim de aliis speciebus. Item et in tertio, quia in propositione universali ita se habet subiectum ad predicatum quod posito subiecto ponitur predicatum et illo remoto removetur predicatum; sic autem non se habent animal et homo, quia licet posito homine ponatur animal, tamen destructo homine non destruitur animal.
Et ideo dico quod ista propositio 'homo est animal' non est universalis universalitate determinata in hoc libro. Ex quo apparet cuiusmodi conditio est ipsum universale; si enim universale non est nisi convertibilitas predicati cum subiecto, universale vel erit conditio predicati vel predicati in comparatione ad subiectum, et istud ultimum est magis verum.

Ad rationes.

Ad 1.1 Ad primam, cum arguitur 'secundum Themistium genus universaliter predicatur etc.', dico quod genus ita universaliter predicatur de suis speciebus, non sic primo quod prius homini quam asino, sed propter hoc primo, quia non inest homini per aliquod prius. Sed hoc non sufficit ad hoc quod sit universale ⟨universalitate determinata in hoc libro⟩, sed requiritur quod predicatum sit accidens subiecti et sit convertibile cum ipso.

---

38 in^1] *s. l.* M

subject's proper attribute in every case, since the subject and its attribute are not as the exceeding and the exceeded, but in which one is found and also the other, and vice versa. And this is what the Philosopher means here by 'universal'. Accordingly, he says that that proposition is universal in which the predicate inheres in the subject through the nature of the subject with which it is converted, in such a way that the predicate is not in the subject through anything other than the subject.

In response to the question, then, it should be replied that this proposition, 'A human being is an animal', in one way agrees with the universal, but disagrees in three ways. For it agrees with the universal in this, that animal inheres in human being universally, *per se*, and primarily, as was shown. But in three ways it disagrees. First, since in a proposition in which the predicate inheres primarily in the subject it does not inhere in another unless this is through the subject's nature, as 'primarily' is understood here. 'Animal', however, does not agree with 'human being' in this way, indeed, it agrees with 'donkey' to the same extent as with 'human being'. And in another way, too, it disagrees, since in a universal proposition the predicate is the proper attribute of the subject, and not like the exceeding and the exceeded. But 'animal' and 'human being' are not related thus, for 'animal' is not a proper attribute of a human being, nor convertible with 'human being', for it is said of other species. And again in a third way, since in a universal proposition the subject is related to the predicate in such a way that, with the subject posited, the predicate is posited, and with the subject removed the predicate is removed. But 'animal' and 'human' are not related in this way, for although with 'human being' posited, 'animal' is posited, with 'human being' destroyed, 'animal' is not destroyed. And so I hold that this proposition, 'A human being is an animal', is not universal by the sort of universality treated in this book, from which it is apparent what sort of condition this 'universal' is. For if 'universal' is nothing but the convertibility of predicate and subject, 'universal' will either be a condition of the predicate or of the predicate in relation to the subject, and the last is the more true.

In response to the arguments.

Ad 1.1 To the first, when it is argued, 'According to Themistius the genus is universally predicated' etc., I reply that the genus is universally predicated of its species, not primarily in such a way that it is prior to human being rather than donkey, but rather it is primary because of the fact that it does not inhere in human being through anything prior. But this does not suffice for it to be universal ⟨in the sense at issue here⟩. What is required is that a predicate be an accident of the subject and convertible with it.

Ad 1.2 Ad secundam rationem, concedo maiorem. Ad minorem, dico quod licet predicatum insit subiecto primo ita quod non per aliquid aliud prius, tamen hoc non est absolute primo; ad hoc enim quod predicatum conveniat subiecto primo, proprie loquendo de 'primo', oportet quod non insit sibi per aliquid prius, et ita quod non conveniat aliis sicut illi. Sic autem non inest animal homini primo, quia adeo primo inest animal asino sicut homini et omnibus aliis speciebus eius.

⟨**Questio 36**⟩

Consequenter queritur utrum ad naturam universalis exigatur multitudo suppositorum.

1.1 Et arguitur quod sic: Philosophus[a] dicit quod universale est unum in multis et de multis; et Porphyrius[b] dicit quod universale est aptum natum predicari de pluribus; Themistius[c] dicit super librum *De anima* quod universale est conceptus quidam intellectus sine hypostasi ex omni similitudine collectus, similitudo autem non est nisi in pluribus; quare etc.

1.2 Item, de ratione totius unde totum est habere partes plures, quia secundum Boethium[d] impossibile est totum unde totum in una parte consistere; sed universale unde tale est quoddam totum; ergo etc.

1.3 Item, per Philosophum[e] I *Celi et mundi*, in omnibus habentibus formam in materia necesse est esse multitudinem suppositorum in actu; sed in omnibus universalibus habetur forma in materia; ergo etc.

2.1 Oppositum arguitur: Philosophus[f] dicit VII *Metaphysice* quod sol et luna sunt universalia, et si applicetur eis definitio erit universalis; sed sol et luna non habent plura supposita; ergo etc.

---

[a] Cf. Aristoteles, *Int.*, 7, 17 a 37–38; *Met.*, VII, 13, 1038 b 10–11   [b] Cf. *Porphyrii Introductio in Aristotelis Categorias a Boethio Translata* (ed. Busse, pp. 26–27)   [c] Themistius, *In De anima*, V (ed. Verbeke, p. 216)   [d] Boethius, *De topicis differentiis* (PL 44, p. 1188 B)   [e] Aristoteles, *DC*, I, 9, 277 b 29–278 a 21   [f] Aristoteles, *Met.*, VII, 15, 1040 a 28–1040 b 1

---

69 omnibus]   hominibus *ante corr.* M   13 forma]   materia M   ‖ materia]   forma M

Ad 1.2 As for the second argument, I grant the major premise. I reply to the minor that although the predicate inheres in its subject primarily in such a way that it is not in it through anything prior, still this is not primary in the absolute sense. For if the predicate is to agree with the subject primarily, strictly speaking, it is necessary that it not inhere in it through anything prior, and this in such a way that it does not agree with any other in the same way. But 'animal' does not inhere in a human being primarily, since 'animal' inheres in a donkey, and in all its other species, just as much as it does in a human being.

Question 36.

Next is it is asked whether a multitude of supposita is needed for the nature of a universal.

1.1 It is argued that it is, for the Philosopher says that a universal is one in many and from many. And Porphyry says that a universal is suited by nature to be predicated of many. Themistius says, in *De anima*, that a universal is a concept of the intellect, collected from every similar instance, without the hypostasis. But being similar occurs only where there are many things. Therefore etc.

1.2 Furthermore, the account a whole, considered as a whole, is to have many parts, for according to Boethius it is impossible for a whole considered as a whole to consist in one part. But a universal considered as such is a kind of whole. Therefore etc.

1.3 Furthermore, according to the Philosopher in *De caelo et mundo*, in everything having a form in matter it is necessary that there be a multitude of supposita in act, but in every universal there is form in matter; therefore etc.

2.1 On the other hand, it is argued that the Philosopher says in *Metaphysics* VII that sun and moon are universals, and if a definition is applied to them it will be universal. But sun and moon do not have many supposita. Therefore etc.

2.2 Item, ad naturam illius non exigitur multitudo suppositorum, quod potest salvari in qualibet parte; sed universale unde tale potest salvari ad quamlibet sui partem; ergo ad naturam universalis unde tale non requiritur multitudo individuorum in actu.

3 Ad hoc est intelligendum quod in entibus inveniuntur quattuor genera universalium, duo in generabilibus et duo in sempiternis. In generabilibus inveniuntur aliqua generata per propagationem, aliqua per putrefactionem. In generatis per propagationem oportet esse duo supposita ad hoc quod natura rei salvetur, scilicet masculum et femellam, et unum erit agens, reliquum patiens. In generatis per putrefactionem non est necesse esse plura supposita in actu, quia forma inducitur in illis per virtutem corporum superiorum; unde dicit Philosophus[a] quod pater generatorum per putrefactionem est sol. Causa quare in generatis per propagationem universale requirit necessario plura secundum actum et quare oportet illa esse distincta secundum diversitatem sexus est illa, quia ad hoc ut natura speciei salvetur in talibus oportet ut fiat generatio individuorum, et ideo dicit Philosophus[b] quod in ipsis entibus data est generatio propter continuationem speciei. Generatio autem individuorum | non posset esse nisi per actionem ⟨duorum⟩ quorum unum esset patiens et reliquum agens, et propter hoc, licet mulier non sit intenta a natura particulari speciei, est tamen intenta a natura universalis, cuius est continuare species in esse.

M 106rb

Universale autem in generatis per putrefactionem non requirit plura supposita in actu, sed in potentia solum. Cuius ratio est, quia in ipsis nullum invenitur individuum quod possit semper manere, immo nec aliquod in sphera activorum; nunc autem species in talibus non potest salvari nisi per esse alicuius individui; ideo oportet ponere in ipsis successionem individuorum, ut salvetur natura speciei, ita ut corrupto uno generetur aliud. Propter quod universale repertum in ipsis, licet non habeat multa supposita secundum actum et simul, oportet ut habeat multa secundum potentiam necessario.

In sempiternis etiam inveniuntur duo genera entium: unum est substantia sempiterna corporea, aliud est substantia sempiterna incorporea, quod

---

[a] Aristoteles, *GA*, II, 6, 743 a 32–35   [b] Aristoteles, *HA*, V, 1, 539 a 16–25

17 illius] dicitur multitudo *add. sed del.* M   27 quia] per *ante corr.* M   29 propagationem] putrefactionem M   32 generatio] genus M || et] ita *add. sed del.* M   33–34 continuationem] constitutionem *ante corr.* M   36 particulari] *s. l.* M   44 aliud] et e converso *add. sed del.* M   47 entium] accidentium M

2.2 Furthermore, a multitude of supposita is not required for a nature that can be preserved in any part. But a universal as such can be preserved with respect to any of its parts, therefore a multitude of individuals in act is not required for the nature of a universal as such.

3 In response to this, it should be understood that four genera of universals are found in beings, two in generable beings and two in sempiternal beings. In generable things some are found that are generable through propagation, some through putrefaction. Among those generated through propagation there must be several supposita for the nature of the thing to be preserved, namely a male and a female, and one will act and the other receive action. Among those generated by putrefaction it is not necessary that there be many supposita in act, for the form is brought into these by the power of a superior body. Hence the Philosopher says that the father of things generated through putrefaction is the sun. The reason why the universal among things generated by propagation necessarily requires many in act, and why these must be distinct in sex, is that for the nature of the species to be preserved in such cases it is necessary that there be a generation of individuals, and so the Philosopher says that in these beings there is generation for the continuation of the species. But the generation of individuals can occur only through the action of two things of which one will act while the other receives the action. And because of this, even though the female is not intended by the particular nature of the species, yet it is intended by the nature of the universal, to which it belongs to continue the species in being.

But the universal, in the case of things generated by putrefaction, does not require many supposita in act, but only in potentiality. The reason is that, among these, no individual is found that can always remain, nor indeed is any such individual found in the sphere of active things. But a species of this sort cannot be preserved except through the being some individual. Therefore it is necessary to posit in these species a succession of individuals, so that the nature of the species is preserved in such a way that when one is corrupted another is generated. Thus the universal found in these, although it does not have unchanging supposita in act, yet at the same time it necessarily has many in potentiality.

Among sempiternal things, too, there are two genera of beings. One is corporeal sempiternal substance and the other is incorporeal sempiternal

est absolutum a motu et a materia. In substantiis corporeis sempiternis non sunt plura individua in actu nec in potentia, quia si ibi esset potentia ad plura, illa potentia ordinaretur ad impossibile; ideo non possunt esse plures soles et plures lune. Et ratio huius est, quia in istis in quibus est pluralitas individuorum causa multiplicationis est quia tota materia debita illi speciei non occupatur in uno individuo, et iterum, quia individuum non est tale ut forma unius individui possit satiare appetitum materie speciei. Nunc autem in sempiternis ita est quod materia que debetur toti speciei occupatur in uno individuo, et sub forma eius, et forma unius individui satiat totum appetitum materie, sicut est de sole. Unde preter hunc solem non invenitur aliquid quod sit in potentia ad formam solis, et ideo tota materia occupatur in uno individuo, et sua forma satiat appetitum materie. Iterum, cum est sub ⟨sua⟩ forma non appetit esse sub alia, nec est ibi admixta privatio, et similiter est de mundo et luna et aliis, et ideo in talibus non est multiplicatio individuorum secundum actum nec secundum potentiam. In quibusdam autem aliis est multiplicatio secundum actum et in quibusdam secundum potentiam, ut visum est.

Sed tu queres: quare est ita ⟨quod⟩ in corporibus ⟨sempiternis⟩ unum individuum occupat totam materiam debitam uni speciei et non hic? Dico quod est, quia approximant magis illi quod est maxime unum et maxime perfectum, et ideo providit natura ut ipsa sint incorruptibilia et secundum speciem et secundum individuum, et quod ista inferiora sint corruptibilia ad minus secundum individuum. Licet tamen in sempiternis non inveniatur multitudo suppositorum, nec secundum actum nec secundum potentiam, tamen bene forte invenitur ibi multitudo suppositorum secundum aptitudinem, quia forme solum non repugnat esse in pluribus, et ideo dicimus esse aptum natum quod non habet repugnantiam ad illud. Et non sequitur ulterius: habet aptitudinem ad hoc, ergo habet potentiam, sicut patet de ceco, quia cecus est aptus natus ad videndum, non tamen habet potentiam ad videndum, quia a privatione etc.[a]

In substantiis autem sempiternis incorporeis non invenitur multitudo suppositorum nec secundum actum nec secundum potentiam nec secundum aptitudinem, quia omnis multitudo suppositorum est ex divisione materie, et divisio materie ex partibilitate quantitatis, et ideo, quia in ipsis

---

[a] Cf. Aristoteles, *Cat.*, 10, 13 a 31–33

55 individui] speciei *ante corr.* M   67 materiam] naturam M   ‖ hic] hoc M   71 corruptibilia] incorruptibilia *ante corr.* M   80 actum] aptum *ante corr.* M

substance, which is unrestricted by motion and matter. In the case of corporeal sempiternal substances there are not many individuals in act, nor in potential, for if there were potential there for many, that potential would be ordered to something impossible. Therefore, there cannot be many suns and many moons, and the reason for this is that in those in which there is a plurality of individuals the cause of multiplication is that all the matter devoted to that species is not taken up in one individual, and again because the individual is not such that the form of one individual can satisfy the appetite of the matter for the species. Now in sempiternal things all the matter devoted to the species is taken up in one individual, and under its form. And the form of one individual satisfies the whole appetite of the matter, as occurs in the sun. So aside from the sun, nothing is found that is in potentiality to the form of the sun, and so all the matter is taken up in one individual, and its form satisfies the appetite of the matter. Furthermore, when it is under its own form, it does not have an appetite to be under any other, nor is any privation mixed in there. And the same is true of the earth, the moon, and so on. And so there is no multiplication of individuals among these, either according to act or potentiality. But in certain other cases there is multiplication according to act, and in yet others according to potentiality, as we have seen.

But you will ask: Why is it the case that, in sempiternal bodies, one individual takes up all the requisite matter in the one species, whereas this is not the case here ⟨on earth⟩? I say it is because those approximate most closely to that which is most one and most perfect, and so nature provides that those are incorruptible both in species and in the individual, and that lower things are corruptible, at least according to the individual. But even though in sempiternal things a multitude of supposita is not found either according to act or according to potentiality, yet it may well be that a multitude of supposita is still to be found according to aptitude, for being in many is not contradictory to form alone, and therefore we say that that is suited by nature because it is not contradictory to it. And it does not follow, then, that it has aptitude to this and therefore it has potentiality, as is clear in the case of someone blind, for someone who is blind is suited by nature to seeing, but does not have the potentiality to see, for due to his privation etc.

But in incorporeal sempiternal substances a multitude of supposita is not found according to act, nor according to potentiality, nor according to aptitude, for every multitude of supposita is from a division of matter, and a division of matter arises from the ability of quantity to have parts, and

non est materia quanta, ideo non invenitur in ipsis multitudo suppositorum aliquo modo, secundum Philosophum,[a] licet forte oppositum sit tenendum secundum veritatem. Unde dicit Rabin Moyses[b] quod in ipsis non invenitur aliqua distinctio nisi illa que est inter causam et causatum. Et huic consonat auctor[i] *De causis*,[c] dicens quod intelligentia scit res supra se quoniam est causata ab illis et scit res sub se quoniam est causa ipsarum, et hec distinctio solum est cause et causati.

Sed tu queres: quomodo sunt iste universales? Dico quod ex hoc quod sunt actu intellecte sunt universales. Unde dicit Philosophus[d] III *De anima* quod omne separatum est actu intelligens et intellectum, et ex hoc universales sunt in actu. Unde valde longe est ibi ratio universalitatis et in istis inferioribus.

Sed adhuc non videtur dissolvi questio. Questio enim est utrum de ratione universalis unde universale est sit quod habeat multitudinem suppositorum. Forte de ratione universalis unde universale est quod habeat multitudinem suppositorum secundum aptitudinem, sed non secundum actum nec secundum potentiam, quia ex hoc est aliquid universale, quia est actu absolutum preter conditiones individuantes; talis autem absolutio possibilis est si non reperiantur plura nec secundum actum nec secundum potentiam, unde possum intelligere substantiam solis preter quantitatem et preter omnia talia accidentia. Intellectus absolutio facit rationem universalis, quare ad rationem universalis requiritur solum aptitudo ad plura. Et si tu dicas: in substantiis separatis non est aptitudo ad plura, dico quod alia est ratio universalis in ipsis |et in aliis, ut dictum est.

M 106va

Ad rationes.

Ad 1.1 Ad primam, dico quod omnes definitiones ille bene concludunt quod de ratione universalis sit aptitudo ad multa. Unde alibi Philosophus[e] dicit quod universale est aptum natum predicari de pluribus. Et si tu dicas: cui repugnat actus ei repugnat aptitudo, ergo cum aliquibus repugnat actus ad plura, ei repugnat aptitudo ad plura, dico quod maior falsa est, quia non omnis aptitudo ordinatur ad actum.

---

[a] Aristoteles, *Met.*, XII, 8, 1074 b 31–39 [b] Cf. Maimonides, *Dux Perplexorum*, II, 2 (1520, f. 40v) [c] *Liber de causis*, VII (VIII) 72 (ed. Pattin, p. 152) [d] Aristoteles, *De an.*, III, 5, 430 a 18–25 [e] Aristoteles, *Int.*, 7, 17 a 38–b1; *Met.*, VII, 13, 1038 b 10–11

---

87 huic] hinc M 89 ipsarum] ipsorum M 91 intellecte] multiplices M 92 hoc] est *add. sed. exp.* M 99 actum] aptum M 103 absolutio] absolutus M 111 repugnat[2]] potentia *add. sed exp.* M

so, since there is no quantifiable matter in these, neither is a multitude of supposita found in them in any way, according to the Philosopher, although perhaps the opposite is in truth to be maintained. Hence Rabbi Moses says that in these no distinction is found except that between cause and caused. And the author of the *De causis* agrees with this, saying that an intelligence knows things above itself because it is caused by them, and it knows things below itself because it is the cause of them, this distinction being only between causes and the caused.

But you will ask in what way these are universals? I hold that they are universal because they are understood in act. Hence the Philosopher says in *De anima* III that everything separated is in act understanding and understood, and so they are universals in act. Hence the account of universality in these and in inferior things is very far apart.

But this still does not seem to solve the question, for the question is whether it belongs to the account of universal as universal to have a multitude of supposita. Now perhaps it belongs to the account of universal as universal to have a multiplicity of supposita according to aptitude, but not according to act nor to potentiality, for something is universal due to this, that it is in act absolutely, aside from conditions of individuation. But such absoluteness is possible if many are not found either according to act or potentiality, and so I can understand substance alone aside from quantity and all such accidents. It is absoluteness of understanding that yields the account of universal. Therefore only aptitude toward many things is required for the account of universal. And if you say that there is no aptitude for many in separate substances, I reply that the account of universal is distinct in these cases and in others, as was said.

In response to the arguments.

Ad 1.1 To the first I reply that all these definitions conclude well enough that concerning the account of universal there is an aptitude to many. Whence the Philosopher says elsewhere it is apt by nature to be predicated of many. And if you reply that to that to which the act is repugnant the aptitude is also repugnant, therefore when the act for many is repugnant to something the aptitude for many is repugnant to it, I answer that the major premise is false. For not every aptitude is ordered to an act.

Ad 1.2 Ad secundum, dicendum quod sicut est totum sic habet partes: ⟨si secundum potentiam, habet partes⟩ secundum potentiam, et si secundum actum, habet partes secundum actum, et si secundum aptitudinem, habet partes secundum aptitudinem. Et cum dicitur 'universale est totum', verum est secundum aptitudinem, ut dictum est.

Ad 1.3 Ad aliud, cum arguitur: 'in habentibus formam in materia e⟨s⟩t multitudo individuorum sub una specie per materiam', dicendum quod si maior sumatur universaliter, falsa est, quia quedam habent formam in materia, que materia est in potentia ad formam et non ad privationem forme, et in talibus non requiritur multitudo individuorum per materiam sub una specie, et ideo in talibus poterit manere unum et idem secundum numerum sempiternaliter. Alia autem habent formam in materia que est in potentia ad formam et ad privationem forme, et in istis necessario requiritur multiplicatio individuorum secundum materiam sub una specie, cum unum secundum numerum non possit sempiternaliter manere.

⟨Questio 37⟩

Consequenter queritur utrum genus requirat multitudinem specierum.

1.1 Et arguitur quod non, quia sicut se habet species ad individuum, sic genus ad speciem,[a] quia utrobique est comparatio totius universalis ad suam partem; sed species non requirit multitudinem individuorum in actu; ergo nec genus multitudinem specierum.

1.2 Item, differt totum integrale et totum universale, sicut vult Philosophus[b] V *Metaphysice*, quia totum universale habet partes quarum quelibet est suum totum, totum autem integrale habet partes quarum nulla est suum totum, sicut patet de domo; sed genus comparatur ad species sicut totum universale ad suas partes; ergo ad naturam generis etc.

2 Oppositum arguitur: de ratione totius est habere partes; sed genus est quoddam totum; ergo ad naturam generis requiritur multitudo specierum.

3 Intelligendum est quod in genere est duo considerare, scilicet suam essentiam et suum esse. Si consideretur secundum suam essentiam, non

---

[a] Cf. Aristoteles, *Cat.*, 5, 2 b 18–20   [b] Aristoteles, *Met.*, V, 26, 1023 b 26

---

120 multitudo] et *add. sed forte del.* M || individuorum] secundum speciem *add. sed exp.* M   14 si] requiretur *add. sed. del.* M

Ad 1.2 It should be replied to the second that as it is a whole, so it has parts. If it is a whole in potentiality, then it has parts in potentiality; if it is a whole in act then it has parts in act; if it is a whole in aptitude then it has parts in aptitude. And when it is said that a universal is a whole, this is true according to aptitude, as was said.

Ad 1.3 To the other, when it is argued that in those having form in matter there is a multitude of individuals under one species through the matter, it should be replied that if the major premise is taken universally it is false, because some have form in matter, which matter is in potentiality to form and not to privation of form. In such cases, a multitude of individuals through matter under one species is not required, and in such, one and the same in number can remain sempiternally. But others have a form in matter by which it is in potentiality to form and to privation of form, and in these a multiplication of individuals according to matter under one species is required, since numerically one thing cannot remain sempiternally.

Question 37.

Next it is asked whether a genus requires a multitude of species.

1.1 It is argued that it does not, since just as the species is related to the individual, so the genus to the species, for in both cases there is a relation of the universal whole to its part. But a species does not require a multitude of individuals in act, therefore neither does a genus.

1.2 Furthermore, an integral whole differs from a universal whole, as the Philosopher holds in *Metaphysics* V, since the universal whole has parts of which every one is its own whole. But an integral whole has parts of which none is its own whole, as is obvious in the case of a house. But genus is related to species as the universal whole to its parts, therefore the nature of a genus does not require a multitude of species.

2 On the other hand, it is argued that it belongs to the account of a whole to have parts, but genus is a certain whole, therefore a multitude of species is required for the nature of a genus.

3 It should be understood that there are two things to consider in genus, namely its essence and its being. If it is considered according to its essence

requirit multitudinem specierum, sed quod plus est, ad essentiam eius nulla species pertinet, nec una nec plures. Et ⟨ratio⟩ huius est, quia genus contrahitur ad species suas per differentias contrarias; sed nulla differentia est de essentia generis, quia genus se habet respectu differentiarum sicut materia respectu forme et subiectum respectu accidentium, sed forma non est de essentia materie nec accidentia de essentia subiecti, ergo nec differentie sunt de essentia generis; et cum genus determinetur ad species per differentias, species erunt extra rationem generis, et una et plures.

Sed aliquis diceret: genus secundum suam essentiam consideratum nihil aliud est quam omnes sue species, quomodo ergo non sunt species de eius essentia? Dico quod in genere reperitur unitas, et non reperitur unitas secundum formam sed unitas secundum intellectum et rationem. Licet tunc unum quod est unum secundum formam possit plurificari per diversa numero, tamen impossibile est quod unum et idem secundum formam diversificetur per diversa specie, quia tunc unum essentialiter esset plura essentialiter. Et ideo, cum nos ponamus genus plurificari per plures species diversas, manifestum est quod genus non est unum secundum formam et essentiam, ita quod nomine generis importantur plures res [vel due] et iste res plures sub uno conceptu communi conveniunt apud intellectum, qui quidem intellectus attribuit eis rationem generis. Et iste conceptus animi sumitur ex aliquo apparenti in re, ita quod per genus importantur diverse essentie sub ratione communi, que quidam ratio communis est in differentibus secundum speciem. Ex quo manifestum est quod ad essentiam generis ut sic non pertinet aliqua specierum; illud enim quod pertinet ad essentiam generis ut sic est aliquid unum in ratione, quod quidem reperitur in pluribus secundum speciem. Sed nulla species, sive una sive plures, considerata in propriis naturis potest habere talem rationem communem per quam possit aliqua species reperiri in diversis secundum speciem, et ideo ad rationem generis ut sic neque pertinet una species neque plures species.

Verumtamen species omnes habent rationem communem in genere, sed sic considerare speciem est considerare naturam generis et non speciei. Si autem consideratur quantum ad suum esse, hoc potest esse duobus modis: aut hoc erit secundum esse quod habet in anima aut secundum esse quod habet in re extra. Primo modo non requirit plures species; sicut enim ad essentiam generis non pertinet una species neque plures, sic neque ad intellectum essentialem generis, quia eadem est dispositio rei in entitate et veritate. Si consideretur secundum esse quod habet in re extra, requirit multitudinem specierum, non actu sed potentia, nam cum | genus

M 106vb

---

17 per] ad M   20 nec²] esse *add. sed del.* M   31 non est] *inv.* M   35 animi] animus M
37 differentibus] communis *add. sed del.* M

then it does not require a multitude of species, and what is more, no species pertains to its essence, neither one nor many. And the reason for this is that a genus contracts to its species through contrary differences. But no difference belongs to the essence of a genus, since genus is related to differences as matter is to form and subject to accidents, but form is not of the essence of matter, nor accidents of the essence of subject, therefore neither are differences of the essence of genus. And since a genus is determined to its species through their differences, a species will be outside the account of the genus, both one and many.

But someone might say that genus considered according to its essence is nothing other than every one of its species. How is it, then, that its species are not of its essence? I reply that unity is found in the genus, and not unity in form, but unity in concept and account. Then, even though what is one in form can be made many through diversity in number, still it is impossible that one and the same form be made diverse through diverse species, for then one thing essentially would be many essentially. And, therefore, since we assume the genus is made many through many different species, it is obvious that genus is not one in form and essence, so that by 'genus' many things are conveyed, and these many things agree under one common concept in the intellect, which attributes the account of the genus to them. And this concept of the soul is taken from something appearing in reality in such a way that different essences are conveyed through the genus under a common account that is common to things different in species. And from this it is obvious that a given species does not pertain to the essence of a genus as such. For what pertains to the essence of a genus as such is something one in account that is found in many in species. No species, however, whether one or many considered in their proper natures, can have such a common account through which species can be found in things that are diverse in species. And so neither one nor several species pertain to the account of the genus as such.

Now, every species has a common account in the genus, but to consider a species in this way is to consider the nature of the genus and not the nature of the species. But if it is considered in its own being, this can occur in two ways, according to either the being it has in the soul or the being it has in reality outside the soul. In the first way, many species are not required. For just as neither one species nor many pertain to the essence of the genus, neither do they pertain to the essential concept of the genus, for the disposition of a thing is the same in being and truth. If it is considered according to the being it has in the thing outside, it requires

sit natum dividi per differentias oppositas, ipsum genus de necessitate existit sub altera oppositarum differentiarum; subiectum enim existens sub uno contrariorum est in potentia ut sit sub alio. Et cum differentie sint cause formales respectu specierum, simili modo dicendum est de genere respectu specierum, et ideo genus quantum est ad esse extra requirit multitudinem specierum in potentia. Sic igitur patet qualiter ad naturam generis requiritur multitudo specierum et qualiter non.

Modo dicendum est ad rationes.

Ad 1.1 Ad primam, dico quod quodam⟨modo⟩ est similitudo: similitudo est, quia sicut species potest salvari in uno individuo, sic genus in una specie, et sicut species predicatur in quid de individuo, sic genus de specie. Sed dissimilitudo est, quia genus [per differentias] determinatur ad species per differentias que accipi⟨un⟩tur ex parte forme habentis contrarium, et ideo necessario, cum genus sit sub una specie, requirit quod possit esse sub alia, quia si unum contrariorum est in natura, et reliquum poterit esse de necessitate. Et ideo cum genus per differentias determinetur ad species, necessario habet potentiam ut sit sub alia. Sed species determinatur ad individuum per materiam, et materie ut sic nihil contrariatur, et ita ratione contrarietatis non est causa quare species requirat multitudinem individuorum, immo contingit reperire tale individuum quod semper manet, quia, cum materia sit sub illo individuo, non est in potentia ut sit sub alio; talia autem sunt sol et luna, ut visum est.

Ad 1.2 Ad aliam rationem, cum dicitur 'genus est totum universale', concedo, et ideo quelibet species est ipsum quod est genus realiter. Et cum dicitur 'ergo sufficienter potest salvari in una specie', verum est secundum actum, et causa dicta est.

---

60 qualiter] dicendum *add. sed del.* M   75 luna] ideo etc. *add. sed del.* M   76 genus] erit *add. sed del.* M

a multitude of species not in act, but in potentiality. For since a genus is suited to be divided through opposed differences, the same genus of necessity exists under one or the other of the opposed differences. For the subject existing under one of the contraries is in potentiality to exist under the other. And since differences are formal causes with respect to species, it must be held in a similar way of the genus with respect to the species. And so genus, as far as its being outside the soul is concerned, requires a multitude of species in potentiality. Thus it is clear in what way a multitude of species is required for the nature of the genus and in what way it is not.

A reply should now be made to the arguments.

Ad 1.1 To the first, I answer that in a certain way there is a similarity, because just as the species can be preserved in one individual, so the genus can be preserved in one species; and just as the species is predicated essentially (*in quid*) of an individual, so the genus is predicated of the species. But there is a dissimilarity as well, since the genus is determined to its species through differences that are received on the part of a form having a contrary. And so it is necessary that since a genus is under one species, it is required that it can be under another, for if one of the contraries exist in nature, necessarily the other can exist. And so, since a genus is determined to its species through differences, it necessarily has the potentiality to exist under another. But a species is determined to the individual through matter, and nothing is contrary to matter as such, and so, by the account of contrariety there is no cause for why a species requires a multitude of individuals. Instead, it is contingent whether an individual is found that always remains, since when the matter is under one individual it is not potentially existent under another, as with the sun and the moon, as we have seen.

Ad 1.2 In response to the other argument, when it is said, 'genus is a universal whole', I grant this. And so every species is that which is the genus in reality, etc. It is said, 'therefore it can be sufficiently preserved in one species', and this is true in act, and the cause has been given.

⟨Questio 38⟩

Consequenter queritur utrum cognoscens omnem particularem triangulum habere tres cognoscat triangulum habere tres.

1 Et arguitur quod sic, quia quecumque se habent sicut relativa, cognito uno cognoscitur reliquum, quia cognito uno relativorum etc.; sed omnes particulares trianguli convertuntur cum triangulo universaliter dicto; ideo etc.

2 Oppositum dicit Philosophus;[a] dicit enim quod quamvis aliquis demonstret de gradato et ysochele habere tres, non demonstrat de triangulo habere tres, nisi sophistico modo.

3 Dico ad hoc quod cognoscens omnem triangulum particularem habere tres ⟨non cognoscit triangulum habere tres⟩, loquendo de scire simpliciter, quia ad hoc quod aliquis sciat passionem de subiecto oportet quod sciat per propriam rationem subiecti et non per aliquod accidens; sed ratio omnium particularium triangulorum accidit rationi trianguli in communi, immo ratio eorum est extra rationem trianguli communis; ideo cognoscens omnem triangulum particularem habere tres non cognoscit triangulum habere tres per rationem propriam trianguli.

Sed aliquis dicet: Commentator super II *Metaphysice*[b] dicit quod universalia sunt bone mixtionis cum particularibus, fortius enim inherent suis particularibus quam accidentia suis subiectis; sed ex cognitione propriorum accidentium intellectus devenit in cognitionem subiecti; ergo multo fortius ex cognitione particularium devenimus in cognitionem universalis, et ita sciens omnem triangulum particularem habere tres, scit triangulum habere tres.

Verum est quod nullum accidens per se et primo et in virtute propria ducit in cognitionem quidditativam substantie, tamen verum est quod ex operatione aliquando devenimus in cognitionem substantie, quia cum in animali apparet ista operatio que est sentire, et ista operatio non potest reduci in materiam nec dimensiones [sentiret] nec in qualitates primas –

---

[a] Aristoteles, *AnPo*, I, 5, 74 a 26–32 [b] Averroes, *Commentarium magnum in Metaphysicam Aristotelis*, I, 9, comm. 31 (1562, f. 20H)

---

5 trianguli] cognoscuntur *add. sed del.* M   19 particularibus] universalibus M   27 quia] sed M   29 nec¹] in *add. sed exp.* M || qualitates] quantitates M

Question 38.

Next it is asked whether one who cognizes that every particular triangle has three angles equal to two right angles cognizes that triangle has three angles etc.

1 It is argued that he does, since whenever things are related to one another as relatives, when one is cognized the other is cognized, for whenever one of two relatives is cognized, etc. But all particular triangles are convertible with triangle said universally. Therefore etc.

2 The Philosopher says the opposite. For he says that even though someone demonstrates of a scalene and an isosceles triangle that they have three etc., he does not demonstrate of triangle that it has three etc., except sophistically.

3 I reply to this that one who cognizes that every particular triangle has three etc. does not cognize that a triangle has three etc., speaking of knowledge without qualification. This is so because for someone to know the attribute of the subject it is necessary that he know this through the account proper to the subject and not through any accident. But the account of every particular triangle is accidental to the account of triangle in common, indeed, their account is outside the account of triangle in common. Therefore one who cognizes that every particular triangle has three etc. does not cognize that triangle has three etc. through the account proper to triangle.

But someone will claim that the Commentator says on *Metaphysics* II that universals are from a good mixture with particulars, for they inhere more strongly in their particulars than accidents in their subjects. But from the cognition of proper accidents the intellect arrives at a cognition of the subject. All the more, then, from a cognition of its particulars we will arrive at a cognition of the universal, and so one who knows that every particular triangle has three etc. knows that triangle has three etc.

It is true that no accident, in virtue of itself, leads *per se* and primarily to a quidditative cognition of its substance. But it is also true that from its operation we do sometimes arrive at a cognition of its substance. For when the operation of sensing appears in an animal, this operation can be reduced to neither its matter, nor its dimensions, nor its primary

quia tunc ignis sentiret quia est calidus – nec qualitates posteriores, [et] ideo reducimus ipsam in aliquam formam substantialem. Sic ergo ex operatione discurrit intellectus in cognitionem substantie a qua progreditur talis operatio, et non cuiuscumque, nec devenit in cognitionem substantie quocumque modo, sed prout est causa illius operationis.

Habesne adhuc intellectum essentialem substantie? Dico quod non, quia quemadmodum tu intelligis subiectum in comparatione ad aliud subiectum, simili modo ex parte ista dico quod particularia in virtute propria non ducunt in cognitionem universalis. Tu dices: in virtute cuius? Tunc dico quod in virtute intellectus agentis; intellectus enim agens videns omnia particularia convenire in una natura communi abstrahit ipsam ab omnibus principiis individuantibus et facit ipsum universale. Unde bene dixit Philosophus[a] quod cognoscens omnem particularem triangulum habere tres non cognoscit simpliciter de triangulo quod habeat tres secundum quod triangulus, sed forte secundum accidens.

Ad argumenta.

Ad 1 Ad primum, dico quod relativa sunt dupliciter: sunt enim secundum substantiam et sunt secundum esse; secundum substantiam sicut definitio et definitum, secundum esse sicut universale et particulare. Qui ergo cognoscit distincte unum relativorum ⟨secundum substantiam⟩, distincte cognoscit et reliquum, qui tamen cognoscit unum relativorum secundum esse non oportet quod cognoscat aliud nisi secundum esse. Et ideo, quia omnia particularia in esse sunt idem cum ipso universali et convertibilia cum ipso, sciens omnia particularia scit ipsum universale secundum esse quod habet in re extra, sed hec cognitio ipsius universalis non est cognitio ipsius secundum substantiam suam; immo quia ipsum universale secundum substantiam suam et omnia particularia non convertuntur nec sunt ad invicem relativa, ideo cognoscens omnem triangulum habere tres non cognoscit simpliciter de triangulo secundum substantiam suam, et sic intelligit illud Philosophus.

---

[a] Cf. Aristoteles, *AnPo*, I, 4, 73 b 29

---

31 ideo] credimus *add. sed del.* M 36 quemadmodum] quandocumque M 38 non] *s. l.* M 42 cognoscens] cognoscem M 51 cognoscat] reliquum secundum omne *add. sed del.* M

qualities—since in that case fire would sense that it is hot—nor its posterior qualities. We therefore reduce it to some *substantial* form. In this way, then, the intellect progresses from the operation to a cognition of the substance from which such an operation arises. And not from everything, nor in every way, does one arrive at a cognition of the substance, but only as it is the cause of that operation.

Still, do you not have an essential concept of the substance? I say no, for just as you understand one subject in relation to another subject, similarly in this case I hold that particulars by their own power do not lead to a cognition of the universal. You will say, by the power of what, then, do they lead to the cognition of the universal? I hold that it is by the power of the agent intellect, for the agent intellect, seeing every particular to agree in one common nature, abstracts this nature from every principle of individuation, and produces the universal. Whence the Philosopher was right to say that one who cognizes that every particular triangle has three etc. cognizes without qualification that triangle has three, considered as triangle, although perhaps he does so accidentally.

In reply to the arguments:

Ad 1 To the first I answer that relatives are said in two ways, for they are said according to substance and according to being. According to substance they are said as definition and the defined, and according to being they are said as universal and particular. Whoever, then, distinctly cognizes one relative according to substance distinctly cognizes the other. But whoever cognizes a relative according to being does not necessarily cognize the other except according to being. And so, since all particulars in being are together the same as the universal and convertible with it, one who knows every particular knows the universal according to the being that it has in eternal reality. But this cognition of the universal is not a cognition of it according to its substance. Rather, the universal according to its substance is not convertible with all its particulars taken together, nor are they relatives with respect to one another. So one who cognizes that every triangle has three etc. does not cognize without qualification, concerning triangle according to its substance, that it has three etc. And that is how the Philosopher understood this.

⟨Questio 39⟩

Consequenter queritur utrum | genus acceptum in sua communitate predicetur de specie, puta animal sumptum in sua communitate predicetur de homine. M 107ra

1.1 Et arguitur quod non, quia si animal in sua communitate possit predicari de homine, ergo cui non repugnat animal in sua communitate, ei non repugnat homo; sed asino non repugnat animal in sua communitate; ergo nec homo sibi, ergo hec erit vera: 'homo est asinus'.

1.2 Item, in propositione affirmativa predicatum est idem numero cum subiecto; sed animal acceptum in sua communitate non est idem numero cum homine; ideo etc. Probatio minoris, quia per Philosophum[a] I *Topicorum* unum numero dicitur tripliciter, scilicet unum definitione, unum proprio, unum accidente. Sed animal in sua communitate non est proprium hominis nec accidens; quod non sit definitio, probatio, quia definitio est convertibilis cum definito, tale autem non est animal in sua communitate respectu hominis. Ergo etc.

2 Oppositum arguitur: nihil quod non est acceptum in sua communitate potest predicari de alio sicut genus; sed animal predicatur de homine sicut genus; ergo ut sic accipitur in sua communitate.

3 Intelligendum est quod nomine generis importantur duo, scilicet intentionem generis et rem subiectam intentioni. Et cum dicimus 'predicari de specie' non est hoc intelligendum de predicatione que est intentio generis, quia illud quod predicatur de partibus est aliquid existens in ipsis. Tale autem non est intentio generis, quia illa solum est in anima, quia si esset extra animam, tunc circumscripto intellectu adhuc maneret, quod falsum est; nulla enim intentio logicalis manet circumscripto intellectu. Propter quod cum dicimus 'genus predicari de specie' intelligendum est de re subiecta intentioni.

Sed est intelligendum ulterius quod illud predicat terminus quod significat, et nihil aliud, idem autem significat nomen et definitio; ratio enim, cuius nomen est signum, definitio est, per Aristotelem[b] IV *Metaphysice*, et

---

[a] Aristoteles, *Top.*, I, 7, 103 a 23–31   [b] Aristoteles, *Met.*, IV, 7, 1012 a 24–25

21 que ... generis] *i. m.* M   26 re] *i. m.* M   29 idem] nihil M

Question 39.

Next it is asked whether the genus taken in its commonness is predicated of the species, for instance, if 'animal' taken in its commonness is predicated of a human being.

1.1 It is argued that it is not, since if 'animal' in its commonness can be predicated of human being, then human being is not contrary to whatever animal is not contrary to in its commonness. But 'animal' is not contrary to a donkey in its commonness, so neither is 'human being'. So this will be true, 'A human being is a donkey'.

1.2 Furthermore, in an affirmative proposition the predicate is the same in number with the subject, but 'animal' taken in its commonness is not the same in number with a human being, therefore etc. Proof of the minor premise: according to the Philosopher in *Topics* I, one in number is said in three ways, namely one by definition, one by property, and one by accident. But 'animal' in its commonness is not a property or an accident of a human being. A proof that it is also not one definition is that a definition is convertible with the defined, and animal in its commonness is not convertible with human being, therefore etc.

2 On the other hand, it is argued that nothing that is not taken in its commonness can be predicated of another as its genus, but 'animal' is predicated of a human being as its genus, therefore as such as it is taken in its commonness.

3 It should be understood that two things are conveyed by the name of a genus: namely, the intention of the genus and the thing subject to that intention. And when we say 'predicated of the species', we should not understand that the intention of the genus is predicated, for that which is predicated of its parts is something existing in them. But the intention of the genus is not like this, for it is only in the soul, for if it were outside the soul then, with the intellect set to one side it would remain, which is false. For no logical meaning remains with the intellect restricted. Because of this, when we say of a genus that it is predicated of a species, this must be understood concerning the thing subject to the intention.

Next, it should be understood that a term predicates what it signifies and nothing else. Now name and definition signify the same thing, for the account of which the name is a sign is a definition, according to Aristotle in

ideo illud quod indicat definitio rei terminus subicit et predicat. Et illud quod indicat definitio animalis est hec substantia animata sensibilis. Sic ergo dicendo 'homo est animal' nihil aliud est quam 'homo est substantia animata sensibilis', unde intentio non predicatur, sed res predicatur cui applicabilis est intentio.

Sed ulterius restat: predicaturne animal ut superius, ut commune et ut universale? Dico quod non, quia licet illud quod predicatur fuerit commune, non tamen predicatur ut commune. Cuius ratio est, quia communitas, universalitas et intentio generis non accidunt animali nisi ut animal intellectum est. Et hoc apparet [quia] per Avicennam,[a] qui dicit quod rebus naturalibus secundum duplicem modum essendi duplicia debentur accidentia; habent enim res naturales esse extra animam, et secundum hoc esse debentur eis accidentia talia cuiusmodi sunt album, nigrum, et talia habent etiam esse in anima, et secundum hoc esse debentur eis ista accidentia: esse genus et esse speciem et huiusmodi. Ex hoc patet quod talia accidentia, esse genus vel universale, non conveniunt animali nisi ut intellectum est. Sed constat quod hec predicatio est per accidens: 'homo est animal secundum quod intellectum est', quia predicatum affirmatur de subiecto sub aliquo accidente. Manifestum est autem quod esse genus et esse speciem accidunt animali et homini, quare si genus predicetur de specie per se, non est dicendum quod animal predicetur de homine ut genus est, nec ut commune; quod igitur predicatur ibi est commune, sed non predicatur ut commune est vel genus.

Sed diceret aliquis: illud quod 'animal' significat est aliquid ut indeterminatum est, ⟨quod est⟩ significare illud ut commune, ergo 'animal' significat aliquid ut commune, ergo cum illud quod significat et predicat, et e converso, ergo predicando animal de homine predicat aliquid ut commune. Dico quod quando dicimus quod 'animal' significat aliquid indeterminate, credo quod illud 'indeterminate' nec includitur in significato 'animalis' nec in modo significandi eius, quia natura animalis ut significatur per 'animal' nec determinata nec indeterminata est. Et ratio huius est, quia illud quod convenit aliqui nature secundum se convenit omnibus participantibus naturam illam, sicut patet de habere tres respectu trianguli. Si ergo dicas quod ratio indeterminati convenit animali ut sic significatur,

---

[a] Avicenna latinus, *De Philosophia prima vel scientia divina*, V, 1 (ed. Van Riet, p. 227)

---

31 terminus subicit] *inv. et ordinem restituit* M   38 ratio] *post. corr.* M   46 vel] ut M   55 significare] singulare M   56 cum] et M   ‖ quod] *i. m.* M   60 quia] sed M

Metaphysics IV. And therefore that which a definition of a thing indicates is what the term makes subject and predicates, and that which the definition of 'animal' indicates is this animated sensible substance. Thus, in saying 'a human being is an animal', nothing is said but 'a human being is an animate sensible substance'. Whence the intention is not predicated, but the thing is predicated to which the intention is applicable.

But the question remains: is 'animal' predicated as it is superior and common and universal? I hold that it is not. For although that which is predicated is common, it is not predicated as it is common. The reason for this is that commonness, universality and the intention of genus do not apply to animal except insofar as animal is understood. And this is obvious through Avicenna, who says that natural things have two sorts of accidents in accord with two modes of being. For natural things have being outside the soul, and in accord with this mode of being such accidents belongs to them as white and black. And such things also have being in the soul, and in accord with this mode such accidents belong to them as being a genus, being a species, and so on. From this it is clear that such accidents as being a genus or being universal do not apply to animal except insofar as it is understood. It is clear, moreover, that this predication is *per accidens*, 'A human being is an animal insofar as it is understood', for the predicate is affirmed of the subject under an accident. Now it is obvious that being a genus and being a species are accidents of animal and human being, and so if the genus is predicated of the species *per se*, it is not to be held that animal is predicated of human being insofar as it is a genus, nor inasmuch as it is common. What, then, is predicated there is common, but it is not predicated insofar as it is common, nor insofar as it is a genus.

But someone may reply: what 'animal' signifies is something as it is indeterminate, which is to signify it as it is common; therefore, 'animal' signifies something as it is common; therefore, that which it signifies it also predicates, and conversely; therefore, in predicating 'animal' of a human being one predicates something as it is common. I answer that when we say that 'animal' signifies something indeterminate, I believe that 'indeterminate' is neither included in what 'animal' signifies nor in its mode of signification, since the nature of animal as it is signified through 'animal' is neither determinate nor indeterminate. And the reason for this is that what agrees with a nature does so insofar as it agrees with everything participating in it, as is obvious with having three angles etc. in

ergo convenit homini quod sit indeterminatum ad plures species, cum animal essentialiter de homine predicetur, hoc autem est falsum, cum homo non sit indeterminatum ad plures species; unde 'animal' non significat indeterminate, ita quod determinatio vel indeterminatio cadant in sua ratione.

Sed tu dices, quare dicimus tunc quod 'animal' significat aliquid indeterminate? Dico quod intellectus apprehendens naturam animalis ut sic apprehendit ipsam ut indifferens est ad multas species, et ideo dicimus quod significat indeterminate ad multas species. Propter quod dicendum quod, cum idem predicat terminus quod significat, et ideo, cum 'animal' significet quid commune, cum predicatur de homine predicatur quid commune, non tamen predicatur ut commune, quia 'animal' non significat ut commune; immo si ad illud quod significat solum attendamus, nos dicimus quod non significat nisi esse animalis in composito, sic⟨ut⟩ quodlibet nomen concretum, et manifestum est quod isti rationi accidit universalitas et particularitas.

Ad rationes.

1.1 Ad primam, cum arguitur 'si animal acceptum etc.', dico quod non sequitur, quia animal non est ita commune ad hominem et ad asinum quod illa importet in actu, sed in potentia, quia 'animal', illud quod significat, significat ut homo ab asino non differt. Et istas naturas significant 'homo' et 'asinus' distinctim, et ideo non potest homo de asino predicari, quia significat aliquid ut distinctum ab asino.

Sed tu dices: quecumque uni et eidem sunt eadem, inter se sunt eadem; sed animal dictum de homine est idem numero cum homine, et animal dictum de asino est idem numero cum asino; ergo homo et asinus idem sunt inter se. | Dico quod quecumque uni et eidem determinate sunt eadem, inter se sunt eadem; nunc autem homo et asinus non sunt idem numero in animali determinate sed indeterminate, et ⟨sic⟩ esse idem in numero nihil ⟨aliud⟩ quam esse idem genere; et ideo non oportet quod sint idem inter se.

M 107rb

---

74 significat] terminus *add. sed del.* M   91 determinate] inter se *ante corr.* M

the case a triangle. If, then, you say that the account of the indeterminate agrees with animal as it is thus signified, it then also agrees with human being, for 'human being' is indeterminate with respect to several species, since 'animal' is predicated essentially of a human being. This, however, is false, since 'human being' is not indeterminate with respect to several species. Hence 'animal' does not signify indeterminately in such a way that determination or indetermination falls within its account

But you will answer, 'why do we say, then, that 'animal' signifies something indeterminately?' I reply that the intellect grasping the nature of animal as such grasps it as indifferent to several species. And so we say that it signifies indeterminately in regard of its several species, and because of this it must be held that, since a term predicates what the same term signifies, and since 'animal' signifies something common, when it is predicated of human being something common is predicated. But it is not predicated as common, for 'animal' does not signify as common.[13] Indeed, if we attend only to what it signifies, we hold that it does not signify anything except the being of animal in the composite, as every concrete name does. And so it is obvious that universality and particularity apply accidentally to this account.

In response to the arguments.

Ad 1.1 To the first, when it is argued, 'if "animal" is taken' etc., I reply that it does not follow, since 'animal' is not common to a donkey and a human being as this suggests it is, that is, in act, but rather in potentiality. For that which 'animal' signifies it signifies inasmuch as human being does not differ from donkey. Now, 'human being' and 'donkey' signify these natures as distinct from one another. Therefore 'human being' cannot be predicated of a donkey, for it signifies something as distinct from donkey.

But you will say, 'whatever are the same as one and the same are themselves the same. But the animal said of a human being is the same in number as human being, and the animal said of a donkey is the same in number as donkey, therefore the human being and the donkey are themselves the same'. I reply that whatever are determinately the same as one and same thing are the same as each other. But a human being and a donkey are not the same in number as an animal determinately, but indeterminately, and so being the same in number is nothing than being the same in genus. And so it is not necessary that they be the same as each other.

---

[13] 'Animal' itself is common, in that its meaning does not determine it to a particular species. But its commonality is not included in its definition, since then species that include 'animal' (like 'human') would of themselves be compatible with being a different species, a contradiction. So 'animal' is common, but this is not included as an essential feature of it such that anything that is an animal (or is signified by 'animal') is common.

Ad 1.2 Ad aliud, dicendum quod animal quod predicatur stat in sua communitate, et tamen est idem numero cum homine, et ista stant simul; sed ista non starent simul, si animal predicatum de homine predicaretur de ipso ut commune – staret ut commune, et hoc negatum est; predicatur enim commune, non tamen ut commune, sed solum predicatur de homine sicut illud quod est idem essentialiter et realiter cum homine; immo sic significat et ideo sic predicat.

Ad 1.3 Ad aliud, cum arguitur 'aut est idem definitione etc.', dico quod animal est idem homini definitione pro tanto, quia pertinet ad definitionem et quidditatem hominis.

⟨Questio 40⟩

Consequenter queritur utrum per medium contingens possit sciri conclusio necessaria.

1 Et arguitur quod sic, quia ista est necessaria: 'luna eclipsatur', et tamen medium est contingens, scilicet interpositio terre.

2 Oppositum vult Philosophus.[a]

3 Ad questionem dico quod duplex est medium contingens: unum quod est contingens secundum se et respectu conclusionis, et tale est album respectu huius conclusionis 'habere tres', et per tale medium non potest sciri conclusio necessaria, quia scire est per causam cognoscere, et remota causa removetur effectus; nunc autem remoto medio contingente non removetur conclusio necessaria; ideo illa non potest esse causa conclusionis necessarie. (Aliud est medium quod non est contingens secundum se nec respectu conclusionis, cuiusmodi est habere triangulum extrinsecum etc. respectu huius 'habere tres'.) Aliud est medium quod est contingens secundum se, necessarium tamen respectu conclusionis, et tale est hec 'interpositio terre inter solem et lunam': est contingens secundum se, quia aliquando est et aliquando non est, est tamen necessaria respectu huius conclusionis 'luna eclipsatur', quia posita interpositione terre necessario ponitur eclipsis, et luna remota removetur eclipsis. Et manifestum est quod per tale medium bene scitur conclusio.

[a] Cf. Aristoteles, *AnPo*, I, 6, 74 b 5–75 a 18

98 simul] sed ista non starent simul *add. sed del.* M   100 enim] ut *add. sed exp.* M   7 album] *scripsimus,* album (*sed exp.*) alium M   14 medium] contingens *ante corr.* M

Ad 1.2 To the other argument, it should be replied that the animal that is predicated stands in its commonness, and yet is the same in number as a human being. And these both hold at the same time, but they could not hold at the same time if the 'animal' predicated of a human being were predicated of itself as common. It *would* stand as common, and this is denied. For what is common is predicated, but not *as* it is common. Instead, it is predicated of a human being only as that which is essentially and in reality the same as a human being; indeed so it signifies, and so too does it predicate.

Ad 1.3 In reply to the other, when it is argued, 'either it is the same in definition' etc., I hold that animal is the same as human being in definition inasmuch as it pertains to the definition and quiddity of a human being.

Question 40.

Next it is asked whether one can know a necessary conclusion through a contingent middle term.

1 It is argued that one can, since this is necessary, 'the moon is eclipsed', and yet the middle term is contingent, namely the interposition of the earth.

2 The Philosopher holds the opposite view.

3 I reply to the question that there are two kinds of contingent middle terms. One kind is contingent in itself and with respect to a conclusion, and such is 'that white thing' with respect to this conclusion, that it has three angles etc. And a necessary conclusion cannot be known through such a middle term, for to know is to cognize through the cause, and with the cause removed the effect is removed as well. Now if a contingent middle term is removed the necessary conclusion is not removed, so this middle term cannot be the cause of the necessary conclusion. Another kind of middle term is neither contingent in itself nor with respect to the conclusion, and of such a sort is 'a triangle has an extrinsic angle etc.' with respect to having three angles equal to two right angles. And yet another kind of middle term is contingent in itself but necessary with respect to the conclusion, and such is the interposition of the earth between the sun and the moon. This is contingent in itself, since it sometimes occurs and sometimes does not. But it is necessary with respect to this conclusion,

Sed tu dices: potestne sciri? Loquendo de scire simpliciter, credo quod non, quia quod scitur simpliciter habet esse perpetuum et impossibile aliter se habere. Lunam autem eclipsari non est perpetuum, sed possibile est aliter se habere; contingit enim eclipsem esse et etiam non esse. Et iterum non potest demonstrari de demonstratione simpliciter per illud medium quod est interpositio terre. Et hoc dicit Lincolniensis,[a] quod lunam eclipsari non est necessarium; manifestum enim quod eclipsis particularis non semper est. Iterum, non contingit universale manere si non sit aliquod individuorum, et ideo cum non semper sit aliqua eclipsis particularis, non semper manebit eclipsis universalis.

Et opponit Lincolniensis[b] sibi ipsi: tu dicis quod est universalis et tamen non semper est, nonne sunt universalia perpetua, si debet salvari complete universum? Et respondet quod inconveniens est ponere tale universale non esse, quod quidem universale dicit aliquam naturam subiectam universalitati. Modo eclipsis lune est universale quod non dicit aliquam naturam, sed magis defectum nature; dicit enim magis non ens quam ens. Et talia universalia ponere non esse non est inconveniens. Et ideo dicit quod de eclipsi non est demonstratio nec scientia simpliciter; est ⟨tamen conclusio⟩ simpliciter necessaria de qua possit esse simpliciter scientia.

Ad 1 Ad rationem, cum arguitur 'hoc medium est contingens "interponi terram etc."', verum est secundum se, ⟨sed⟩ respectu huius conclusionis que est 'luna eclipsatur' est necessarium, ita quod posita interpositione terre impossibile est quin fiat eclipsis.

---

[a] Robertus Lincolniensis, *In Posteriorum analyticorum libros*, I, 7 (ed. Rossi, pp. 139–40)
[b] Robertus Lincolniensis, *In Posteriorum analyticorum libros*, I, 7 (ed. Rossi, pp. 139–40)

---

34 quidem] quidam *ante corr.* M   38 simpliciter] *add. spatium vacans fere duodecim litterarum* M

'the moon is eclipsed', for if the interposition of the earth is posited then it is necessary to posit an eclipse, and if the moon is removed then the eclipse is also removed. And it is obvious that we can perfectly well know a conclusion through such a middle term.

But you will object, can it really be known? Speaking of knowing without qualification, I believe that it cannot, since what is known without qualification has perpetual being, and it is impossible for it to stand otherwise. But the moon is not perpetually eclipsed, and it is possible that it be otherwise, for the eclipse can occur, and it can also fail to occur. And again, it cannot be demonstrated with a demonstration without qualification, using that middle term, that is, the interposition of the earth. And Grosseteste says this, that it is not necessarily the case that the moon is eclipsed. For it is obvious that a particular eclipse is not always occurring. Furthermore, it does not happen that the universal remains if there is not some one of its individual instances, and so, since there is not always a particular eclipse, eclipse taken universally will not always remain.

And Grosseteste contradicts himself, you will claim, since he says that it is universal, and yet is not always so. For are not universals perpetual if they are to be preserved as completely universal? And he responds that it is absurd to posit that there is no such nature, because in fact a universal indicates a nature subject to universality. Now a lunar eclipse is a universal which does not indicate any nature, but rather a defect of nature, for it is more a non-being than a being. And to posit that there are no such universals is not absurd, and so he says that there is neither demonstration nor knowledge without qualification of an eclipse. What there can be knowledge of without qualification is a conclusion that is necessary without qualification.

Ad 1 In response to the argument, when it is argued, 'this middle term is contingent, "the earth is interposed" etc.', it is true in itself, but with respect to the conclusion 'the moon is eclipsed', this is necessary in such a way that, given the interposition of the earth, it is impossible that an eclipse should not occur.

⟨Questio 41⟩

Queritur consequenter, quia Philosophus[a] dicit quod corruptio scientie potest contingere vel per corruptionem scientis vel scibilis vel medii, ideo queritur utrum corrupta re corrumpatur scientia.

1.1 Quod non videtur, quia positis causis ponitur effectus, ergo cognitis causis cognoscitur effectus; sed non existente scibili potest sciri causa eius, ut corrupta pluvia potest sciri causa pluvie; quare etc.

1.2 Item, quando alicui rei convenit duplex esse quorum unum non dependet ex alio, corrupta re quantum ad unum, non oportet quod corrumpatur quantum ad aliud; sed esse in re extra et esse in anima sunt duo quorum unum non dependet ab alio, ergo corrupto scibili quantum ad esse extra, non oportet corrumpi quantum ad esse in anima, esse autem in anima est sua cognitio ab anima; quare etc.

2.1 Oppositum vult Philosophus[b] hic et in V *Metaphysice* super ⋯.

2.2 Item, dicit Philosophus[c] in principio quod de subiecto oportet supponere quid est et quia est; quare ipso corrupto quantum ad suum esse, et quantum ad suum scire destruetur.

3 Intelligendum quod ⟨per⟩ scibile potest intelligi illud quod scitur, ut conclusio demonstrationis; differt enim intellectus et scientia, quia scientia est conclusionum, intellectus autem principiorum, et ideo potest per scibile intelligi illud quod scitur. Item, per scibile potest intelligi illud de quo scitur, ut subiectum scientie. Loquendo de scibili primo modo, dico quod corrupto scibili corrumpitur scientia, cuius ratio est, quia tale scibile | non refertur per se ad scientiam, sed scientia ad ipsum. Quedam enim sunt relationes, secundum Commentatorem[d] II *De anima*, que fundantur super substantiam utriusque extremi, ut pater et filius, duplum et dimidium, et quedam super substantiam alterius tantum, et talia sunt

M 107va

---

[a] Cf. Aristoteles, *AnPo*, I, 6, 74 b 34–37   [b] Cf. Aristoteles, *Met.*, V, 5, 1015 b 7–9   [c] Aristoteles, *AnPo*, I, 1, 71 a 13–15   [d] Averroes, *Commentarium magnum in Aristotelis De anima libros*, II, 12, comm. 125–26 (ed. Crawford, pp. 320–22)

---

13 ⋯] *spatium vacans fere sex litterarum* M

Question 41.

Next, since the Philosopher says that the destruction of knowledge is possible through the destruction of either the knower, the knowable or the middle term, it is asked whether, when the thing is destroyed, the knowledge is destroyed.

1.1 It seems that it is not, because when the causes are posited the effect is posited, and therefore if the causes are cognized the effect is cognized. But even if the knowable does not exist its cause can still be known. For instance, one can know the cause of rain when the rain has ceased to exist. Therefore etc.

1.2 Furthermore, when being agrees with some thing in two ways, of which one does not depend on the other, then a thing's being destroyed with respect to one way of being need not mean that it is destroyed with respect to the other. But being in an external thing and being in the soul are two ways of being of which one does not depend on the other. Therefore if some knowable is destroyed according to external being, it need not be destroyed as it is in the soul. But its being in the soul is its cognition by the soul. Therefore etc.

2.1 On the other hand, the Philosopher takes the other side in *Metaphysics* V.

2.2 Furthermore, the Philosopher says at the start of the *Posterior Analytics* that it is necessary to preestablish concerning the subject what-it-is and that-it-is-so. Therefore if it is destroyed with respect to its being it will also be destroyed with respect to its being known.

3 It should be understood that by 'knowable' we can understand here that which is known as the conclusion of a demonstration. For understanding and knowledge differ, since knowledge is of conclusions, whereas understanding of principles. One can in this way understand the knowable as that which is known. Again, by 'knowable' one can understand that which one knows about, as, for instance, the subject of knowledge. Speaking of the knowable in the first way, I hold that if the knowable is destroyed the knowledge is destroyed, the reason for which is that such a knowable does not refer *per se* to knowledge, but rather knowledge refers to it. For there are certain relations, according the Commentator in *De*

mensura et mensurabile, scientia et scibile, quia scientia per se refertur ad scibile, scibile autem non per se refertur ad scientiam. Tunc arguo: illud ad quod aliquid per se refertur, corrupto illo ad quod refertur per se, et ipsum corrumpitur; sed scientia per se refertur ad scibile; quare destructo scibili necessario destruitur scientia.

Item, hoc apparet sic: scientia non est aliud nisi ratio informans intellectum sub relatione ad rem ipsam, unde ad scientiam [namque] requiritur conformitas eius ad rem; sed si corrumpitur scibile quod est conclusio demonstrationis, iam non esset adequatio scientie ad rem; quare si corrupto scibili non corrumpitur scientia, sequitur *** vel quod scientia sit habitus falsus. Et ideo si ponitur triangulum habere tres, destructo isto scibili manifestum est quod destruitur scientia per quam scitur quod triangulus habet tres; scientia enim est ratio existens in intellectu sub conformitate eius ad rem. Unde et, quia ista duo scientia includit, dixerunt quidam quod est in genere ad aliquid et est in genere qualitatis.

Si autem per scibile intelligitur illud subiectum de quo scitur aliquid, sic dico quod corrupto scibili quantum ad aliquid quod convenit sibi secundum se destruitur scientia, corrupto tamen ipso quantum ad esse extra non oportet quod destruatur scientia eius. Primum apparet sic: destructo scibili quantum ad illud quod convenit sibi secundum se, destruitur medium quod est propria ratio illius scibilis. Destructo autem medio quod est ratio propria ipsius scibilis, destruitur scientia, quia ipsa est causa scientie. Destructa autem causa destruitur effectus, quare destructo scibili quantum ad illud quod convenit sibi secundum se destruitur scientia. Et ideo si ponitur quod triangulus destruitur quantum ad quidditatem trianguli, manifestum quod scientia destruetur per quam scitur quod triangulus habet tres. Similiter, destructa quidditate hominis destruitur scientia per quam scitur quod homo est animal rationale.

Destructo tamen scibili quantum ad esse extra non propter hoc destruitur scientia de scibili, cuius ratio est, quia esse et non esse accidunt rei; quod

---

32 ratio] ratione *ante corr.* M   36 ***] *spatium vacans fere unius versus* M   38 quod²] quia M

anima II, that are founded on the substance of each relata (for instance, father and son, double and half), and some on one or the other substance alone (such as measure and measurable, knowledge and knowable), for knowledge is referred *per se* to the knowable, but the knowable is not referred *per se* to knowledge. Then I argue that that to which something is referred *per se*, if that to which it is referred *per se* is destroyed, is itself destroyed. But knowledge is referred *per se* to the knowable, so if the knowable is destroyed then the knowledge is necessarily destroyed.

Furthermore, this is apparent as follows: knowledge is nothing other than an account informing the intellect considered with respect to its relation to the thing itself; hence its conformity to the thing is required for knowledge. But if the knowable that is the conclusion of the demonstration is destroyed, then there would be no proportion of knowledge to thing. Therefore if the knowledge is not destroyed when the knowable is destroyed, it follows that knowledge would be a false habit. So if it is established that a triangle has three etc., then it is clear that the knowledge is destroyed when this knowable is destroyed through which it is known that a triangle has three etc. For knowledge is an account existing in the intellect considered with respect to its conformity to the thing. Hence, since knowledge includes these two aspects, some have held that is in the genus of relation and in the genus of quality.

If, on the other hand, the knowable is understood as the subject of which something is known, then I reply that if the knowable is destroyed (i) as regards something that belongs to it in its own right, knowledge is destroyed. But if (ii) it is destroyed as regards its external being, then it is not necessary that the knowledge of it be destroyed. The first (i) is clear as follows. If the knowable is destroyed as regards that which belongs to it in its own right, then the middle term is destroyed, which is the proper account of that knowable. But if the middle term, which is the proper account of the knowable, is destroyed, then the knowledge is destroyed, for this is the cause of the knowledge, and if the cause is destroyed then the effect is destroyed. So if the knowable is destroyed as regards that which belongs to it in its own right, then the knowledge is destroyed. And so, if it is posited that a triangle is destroyed as regards its quiddity, then it is obvious that that knowledge will be destroyed through which it is known that a triangle has three etc. In the same way, if the quiddity of human being be destroyed, then the knowledge is destroyed through which it is known that a human being is a rational animal.

But (ii) if the knowable is destroyed as regards its external being, then knowledge of the knowable is not on this account destroyed. The reason is

quid enim ex quo quid nec est nec non est, ut probat Avicenna[a] in *Metaphysica* sua, capitulo de universali. Si enim esset, non posset non esse, si non esset, non posset per aliquid esse, nec per aliam causam, cum ex se ipso non sit. Cum ergo de natura intellectus non sit intelligere rem esse vel non esse, ideo potest intellectus intelligere quidditatem absque hoc quod intelligat ipsam esse vel non esse. Tunc arguitur: res solum nata est sciri ex principiis que ingrediuntur suam quidditatem, et ideo corrupta re quantum ad aliquod accidens eius, non oportet quod corrumpatur scientia de ipsa; sed corrupta re quantum ad esse eius extra, corrumpitur solum quantum ad aliquod accidens; quare corrupta re quantum ad esse extra animam, non oportet corrumpi scientiam de ea. Et hoc est quod dicit Egidius[b] super primum *De generatione*, quod res habent triplex esse: in se ipsis, in suis causis, in intellectu, et corrupta re quantum ad esse in se ipsa, non oportet quod corrumpatur quantum ad esse quod habet in suis causis; per tale autem esse habetur scientia de re; et ideo corrupta re quantum ad esse extra animam non corrumpitur necessario scientia de ipsa. Verumtamen ⟨...⟩.

⟨Questio 42⟩

Consequenter queritur utrum corruptibilium possit esse scientia et definitio.

1.1 Et arguitur quod sic, quia rerum naturalium potest esse scientia et definitio, aliter enim de rebus naturalibus scientiam non haberemus; sed res naturales sunt corruptibiles, quia omnes res naturales habent materiam, que est principium corruptionis; ergo etc.

1.2 Item, scientia est habitus semper verus, ergo quorum potest esse habitus semper verus, de eis est scientia; sed de singularibus potest esse habitus semper verus, de Sorte enim possum habere istum habitum 'Sortes est Sortes', et iste est semper verus; ergo etc.

---

[a] Avicenna latinus, *De Philosophia prima vel scientia divina*, V, 1 (ed. Van Riet, p. 277)   [b] Egidius Romanus, *Commentaria in libros de Generatione et corruptione*, ad I, 1, dubium tertium (1505, f. 3rb–va)

60–61 rem ... esse] illa seorsum ibi inesse M   63 sciri] scire *ante corr.* M

that being and non-being are accidental to a thing. For what it is, in virtue of what it is, neither is nor is not, as Avicenna proves in his *Metaphysics* in the chapter on universals. For if it were in virtue of what it is, then it could not fail to exist. And if in virtue of what it is it were not, then it could not exist through anything, nor through any cause other than itself, since it does not exist in virtue of itself. Since, then, it is not of the nature of the intellect to understand a thing as being or not being, the intellect can understand a thing's quiddity without understanding it to be or not to be. Then it is argued that a thing is suited to be known only from principles that involve its quiddity. Therefore if the thing is destroyed as regards one of its accidents, it is not necessary that the knowledge of it be destroyed. But if the thing is destroyed as regards its external being then it is destroyed only as regards some accident. Therefore if a thing is destroyed as regards some being outside the soul, it is not necessary that the knowledge of it be destroyed. And this is what Giles of Rome says on the first book of *De generatione*, that things have being in three ways: in themselves, in their causes, and in the intellect. A thing's being destroyed as regards its being in itself need not mean that it is destroyed as regards its being in its causes. But it is through such being that one has the knowledge of a thing, and so if a thing is destroyed as regards its being outside the soul, the knowledge of it is not necessarily destroyed. On the other hand ⟨...⟩.[14]

Question 42.

Next it is asked whether there can be knowledge and definitions of corruptible things.

1.1 It is argued that there can be, for there can be knowledge and definitions of natural things. For otherwise we would not have knowledge of natural things. But natural things are destructible, since every natural thing has matter, which is the principle of destruction. Therefore etc.

1.2 Furthermore, knowledge is a habit that is always true, and therefore there is knowledge of those things of which there can be a habit that is always true. But there can be a habit that is always true of singulars, for I can have this habit concerning Socrates, 'Socrates is Socrates', and this is always true. Therefore etc.

---

[14] The remainder of the question is not preserved.

2.1 Oppositum patet per Philosophum.

2.2 Et arguitur ratione, quia scientia est semper existentium et perpetuorum; sed singularia non semper sunt, quia per Philosophum[a] VII *Metaphysice* singularibus absentibus a sensu, dubium est utrum sint vel non sint; ergo etc.

3 Ad questionem ⟨dicendum⟩ quod scientia, et definitio, non est rei prout est universalis nec prout est singularis, sed de re absolute. Quod de re non sit scientia secundum quod universalis probatio, quia esse universale accidit rebus; et de nullo est scientia per aliquod quod sibi accidit; ergo etc. Minor patet per Philosophum[b] VII *Metaphysice* contra Platonem; dicit enim quod universale accidit rei. Nec de re est scientia ut particularis est, ⟨quia scientia est⟩ rerum remanentium; res autem ut est particularis corruptibilis est; ergo etc. Sed de re absolute est scientia, quia de eo quod nec generatur nec corrumpitur est scientia; sed quod quid est rei absolute | acceptum nec generatur nec corrumpitur; ergo etc. Probatio minoris, quia omne generatum generatur ex materia sibi proportionali et forma sibi proportionali et agente sibi proportionali; si tunc quod quid est generetur, generatur ex materia et forma sibi proportionali, et talis materia et talis forma non sunt determinate ad hoc vel illud, et ab agente absolute, non hoc vel illo; sed sic ponere generationem est impossibile, ideo etc.

M 107vb

Item, si quod quid est per se generaretur, generato Sorte generaretur omnis homo; sed consequens est falsum. Consequentia patet, quia quod convenit alicui nature secundum se convenit omnibus participantibus naturam illam; sed quod quid est hominis convenit cuilibet supposito hominis; ergo [si] generato Sorte generaretur quilibet homo. Ideo Philosophus[c] dicit VII *Metaphysice* quod quod quid est non generatur absolute, sed quod quid est generatur in hoc. Ipsius ergo rei prout est universalis vel prout est particularis non est definitio [vel] nec scientia, sed rei absolute.

Sed tu dices: tu dicis quod rei prout universalis est non est definitio nec scientia, nec rei prout particularis est. Quare ergo dicit Philosophus magis quod universalium est definitio et scientia quam quod particularium?

[a] Cf. Aristoteles, *Met.*, VII, 15, 1039 b 27–1040 a 7   [b] Cf. Aristoteles, *Met.*, VII, 13, 1038 b 9–11
[c] Cf. Aristoteles, *Met.*, VII, 15, 1039 b 20–27

12–13 perpetuorum] perpect- *sic* M   14 absentibus] ab encitibus *sic* M   16 rei] resi *dub.* M   24 scientia *secunda manu*] essentia *ante corr.* M

2.1 The opposite is clear from the Philosopher.

2.2 It is also argued by reason: Knowledge is of things that always exist and are perpetual, but singulars do not always exist, since, according to the Philosopher in *Metaphysics* VII, when singulars are not present to the senses there is doubt as to whether or not they exist; therefore etc.

3 In response to the question, I answer that knowledge and definition are not of a thing as it is universal, nor as it is singular, but concern the thing absolutely considered. The proof that there is no knowledge of a thing as it is universal is that being universal is accidental to things, and there is knowledge of nothing through what is accidental to it, therefore etc. The minor premise is obvious from the Philosopher, *Metaphysics* VII, against Plato. For he says that being universal is accidental to a thing. Nor is there knowledge of a thing as it is particular, because knowledge is of things that remain, but a thing as it is particular is corruptible; therefore etc. But there is knowledge about a thing without qualification, since there is knowledge about that which neither comes to be nor is destroyed. But the essence of a thing, taken without qualification, is neither generated nor destroyed, therefore etc. Proof of the minor premise: Everything that comes to be comes to be from matter proportionate to it, a form proportionate to it, and an agent proportionate to it. So if its essence comes to be, it comes to be from matter and a form proportionate to it (and such matter and such a form are not determined to this or that), and from an agent taken without qualification (not this one or that one). But to posit such a generation is impossible. Therefore etc.

Furthermore, if an essence comes to be *per se*, then when Socrates comes to be every human being comes to be. But the consequent is false. The inference is clear, since what agrees with some nature according to itself agrees with everything participating in that nature. But the essence of human being agrees with every individual human substance. Therefore when Socrates comes to be, every human being comes to be. Thus the Philosopher says in *Metaphysics* VII that the essence does not come to be without qualification, but the essence comes to be in this particular thing. Therefore, there is no definition or knowledge of a thing as it is universal or as it is particular, but only of a thing without qualification.

But you will answer, 'you hold that there is no definition or knowledge of the thing as it is universal or as it is particular, so why does the Philosopher says that definition and knowledge are more of the universal than of particulars?'

Dicendum est ad primum quod universalium ut universalia sunt non est definitio nec scientia. Verum est tamen quod scientia et definitio sunt habitus quidam intellectuales, et ideo non competunt alicui rei nisi illi que nata est comprehendi ab intellectu, et unumquodque in quantum est comprehensibile ab intellectu habet in tantum rationem universalis, et ideo dicimus quod scientia et definitio est universalium et non particularium. Non tamen est quod quid est rerum secundum quod universales sunt, quia esse universale ut sic est accidens rei, scientia autem non est de aliquo per accidens, quare proprie universalium erit scientia et definitio, non tamen ut universalia sunt.

Ad secundum, dicendum quod in Sorte est duo considerare: [ut] quod quid est et quod quid est hominis ut habet respectum ad accidentia, ut est Sortes hic et nunc per quod quid est hominis. Per quod quid est absolute est Sortes homo et per quod quid est hominis ut habet respectum ad accidentia non est Sortes homo absolute, sed hic homo. Modo Sortes in quantum homo non corrumpitur, sed in quantum hic homo, unde quidditas hominis in Sorte ut absolute consideratur non corrumpitur, sed ipsa ut habet respectum ad accidentia. Et quia illa quidditas hominis in Sorte ut habet respectum ad accidentia est quidditas Sortis secundum quod Sortes, ideo dicimus quod corrupto Sorte corrumpitur quidditas Sortis. Et ideo cum dicitur 'corrupto Sorte corrumpitur quod quid est', verum est, quod quid est Sortis ut Sortes: ut habet respectum ad accidentia. Et quia ea quorum est definitio et scientia oportet quod remaneant, ideo singularis ut Sortis non est scientia neque definitio. Sicut dicendo 'triangulus cupreus habet tres', in quantum cupreus non semper manet ⟨sed⟩ in quantum triangulus, similiter hic: Sortes in quantum hic homo non semper manet, Sortes tamen in quantum homo semper manet. Et quia scientia est rerum secundum quod manent, ideo concedo quod scientia de Sorte ut homo non corrumpitur, sed est necessaria et semper manens.

Ad 1.1 Ad primum, ergo, rerum naturalium est scientia quantum ad suum quod quid est absolute, non unde universales vel particulares. Et cum dicitur 'res naturales sunt corruptibiles', verum est quantum ad eorum esse extra animam.

---

49 quia]   quod quid *add. sed del.* M   70 et]   *s. l. ex corr.* M

It should be replied to the first that there is no definition or knowledge of universals as they are universals, but it is true that knowledge and definition are intellectual habits, and so they are features only of those things that are suited to be grasped by the intellect, and each of them, inasmuch as it is graspable by the intellect, has to that extent the account of a universal. And therefore we say that knowledge and definition are of universals and not of particulars. But the essence is not of things as they are universals, since being a universal as such is accidental to a thing, whereas knowledge is not of something accidental. Properly, therefore, knowledge and definition will be of universals, but not as they are universals.

To the second it should be replied that there are two things to consider in Socrates: the essence and the essence of human being as it is related to accidents, that is, as Socrates is here and now through the essence of human being. According to his essence without qualification, Socrates is a human being, and through the essence of human being as it is related to accidents Socrates is not human being without qualification, but rather he is this human being. Now Socrates is destroyed not inasmuch as he is human being, but rather inasmuch as he is this human being. Hence what is destroyed is not the quiddity of human being in Socrates as it is considered without qualification, but only his quiddity as it is related to accidents. And since the quiddity of human being in Socrates as it is related to accidents is the quiddity of Socrates as Socrates, we say that when Socrates is destroyed the quiddity of Socrates is destroyed. And so when it is said, 'when Socrates is destroyed, his essence is destroyed', this is true of the essence of Socrates as Socrates: as it is related to accidents. And since it is necessary that things of which there is definition and knowledge should remain, there is thus neither knowledge nor definition of the singular considered as Socrates. So, for instance, a copper triangle has three etc., and insofar as it is copper it does not always remain, but insofar as it is triangle it does. In the same way, this Socrates insofar as he is this human being does not always remain, but Socrates insofar as he is human being does. And since knowledge is of things as they remain, I grant that knowledge of Socrates as human being is not corrupted, but it is necessary that it always remains.

Ad 1.1 In response to the first argument, then, there is knowledge of natural things with respect to their essence without qualification, and therefore not as universals or particulars. And when it is said that natural things are destructible, this is true with respect to their being outside the soul.

Ad 1.2 Ad secundum, dicendum quod ibi est fallacia consequentis. Ubicumque enim est scientia est habitus verus, non tamen ubicumque est habitus verus est scientia, priorum enim principiorum est habitus verus, non tamen scientia. Et cum dicitur 'de re corruptibili potest esse habitus semper verus, ut de Sorte quod Sortes est Sortes', dico quod hec non est scientia de Sorte, quia eorum que sunt per se primo modo non est proprie scientia, quia scientia solum est conclusionis, ea autem que sunt in primo modo non habent rationem conclusionis in demonstratione, et ideo ipsorum non est proprie scientia. Unde dico quod istius 'Sortes est Sortes' non est proprie scientia, sed intellectus, nec etiam alicuius propositionis que est per se primo modo.

Sed tu dices: ex dictis tuis sequitur quod particularium sit cognitio nobilior quam scientia, scilicet intellectus. Dico quod aliud est dicere 'Sortes' et 'Sortes est Sortes'; modo istius Sortis non est scientia nec intellectus nec definitio, huius tamen 'Sortes est Sortes' est intellectus et scientia, non tamen proprie scientia, sed intellectus, ut dictum est.

Ad 1.2 In response to the second argument, it should be replied that the fallacy of the consequent occurs there. For wherever there is knowledge there is a true habit, but not everywhere where there is a true habit is there knowledge. For there is a true habit of a first principle, but not knowledge of it. And when it is said there that a habit can always be true of a corruptible thing, for instance, of Socrates that 'Socrates is always Socrates', I reply that this is not knowledge of Socrates, since there is no knowledge strictly speaking of things *per se* in the first mode, for there is knowledge only of the conclusion. But what is *per se* in the first mode does not serve as a conclusion in a demonstration. And therefore there is no knowledge strictly speaking of these, so that I say that there is no knowledge strictly speaking, but only understanding, of 'Socrates is Socrates', nor is there knowledge of any proposition *per se* in the first mode.

But you will reply that from what you have said it follows that there is a cognition of particulars that is more noble than knowledge, namely understanding. I answer that it is one thing to say 'Socrates' and another to say 'Socrates is Socrates'. Now of Socrates there is neither knowledge nor understanding nor definition. There is, however, understanding and knowledge of 'Socrates is Socrates', but strictly speaking there is not knowledge but only understanding, as was said.

⟨Liber II⟩

Questiones sunt equales numero etc. (89 b 24–25)

⟨Questio 43⟩

Circa istum secundum, queritur primo utrum ista propositio sit vera 'questiones sunt equales etc.'. Et hoc est querere utrum numerus questionum attendatur penes numerum scibilium.

1.1 Et videtur quod non, quia si sic, tunc de omni eo esset scientia de quo esset questio; sed hoc est falsum, quia de non ente bene est questio, sed de ipso non est scientia.

1.2 Item, de primis principiis bene est questio; aliqui enim dubitaverunt de primis principiis, et tamen primorum principiorum non est scientia.

1.3 Item, illa que precognoscimus non querimus, quia illud quod precognoscitur non queritur, quia precognitio non est questio; sed que precognoscimus vere scimus; ergo aliqua vere scimus que non querimus, et ita questiones non erunt equales numero his que vere scimus.

1.4 Item, de | substantiis separatis non est questio et tamen de eis est scientia; ergo de aliquibus est scientia de quibus non est questio; quare etc. Maior patet per Philosophum[a] VII *Metaphysice,* qui dicit quod in simplicibus non est questio; sed substantie separate sunt substantie simplices; ergo de ipsis non est questio.   M 108ra

2 Oppositum patet per Philosophum[b] dicentem quod questiones sunt equales numero his que scimus.

3 Dicendum ad hoc quod questiones sunt equales numero his que scimus, quia omnia illa que sciuntur a nobis in parte erant cognita et in

---

[a] Cf. Aristoteles, *Met.,* VII, 17, 1041 a 14–24   [b] Cf. Aristoteles, *AnPo,* II, 2, 90 a 7

---

7 enim] tamen M   21 in parte erant] *iter.* M

Book II

'Questions are equal in number...'

Question 43.

Concerning the second book, first it is asked whether this proposition is true, 'questions are equal in number to whatever truly we know'. And this is to ask whether the number of questions entirely matches the number of knowables.[15]

1.1 It seems that it does not, for if it does, then there would be knowledge concerning everything about which there is a question; but this is false since there is surely a question about non-being, but there is no knowledge about this.

1.2 Furthermore, there is surely a question about first principles. After all, some have raised doubts about first principles. Yet there is no knowledge of first principles.

1.3 Furthermore, we do not ask questions about things we cognize in advance, since what is cognized in advance is not called into question. But what we cognize in advance we truly know; therefore we truly know something that we do not raise questions about, and so there will not be questions equal in number to the things we truly know.

1.4 Furthermore, there is no question concerning separated substances, and yet there is knowledge about these; therefore there is knowledge concerning some things about which there is no question; therefore etc. The major premise is obvious from the Philosopher, in *Metaphysics* VII, who says that there is no question about simple things. Now separated substances are simple substances; therefore there is no question about them.

2 The opposite view is clear from the Philosopher's saying that questions are equal in number to what we know.

3 It should be replied to this that questions are equal in number to what we know, since everything known to us was once in part known and in

---

[15] The vulgate translation of James of Venice reads, 'Quaestiones sunt aequales numero quaecumque vere scimus'. The translation of William of Moerbeke substitutes 'quidem' for 'vere' here. Simon recognizes the potentially misleading interpretation of the vulgate translation and addresses it below.

parte incognita, et omnia talia sunt queribilia. Omnis enim qui querit dubitat et omnis dubitans medius est inter ignorantem et scientem, et ideo dicitur quod in omni questione oportet aliquid supponere et aliquid querere, et ideo omnia que queruntur sunt in parte scita et in parte ignorata. Bene ergo Philosophus iuxta numerum scibilium investigat numerum queribilium.

Sed est intelligendum quod quidam volunt numerum questionum sumi penes numerum scibilium per demonstrationem. Et ratio eorum est, quia omnis questio est de mediatis habentibus causam, quia Philosophus dicit consequenter quod omnis questio est questio medii, et causa et medium sunt idem, et ideo dicunt quod omnis questio est de mediatis et causam habentibus. Sed omnia mediata sunt scibilia per demonstrationem, et ideo dicunt quod solummodo talium scibilium sunt questiones.

Sed si hoc esset verum, nulla esset questio quid est vel si est aliquid: istud non scitur per demonstrationem, immo habetur ante omnem demonstrationem. Si ergo questio solum esset de [pre]scitis per demonstrationem, questiones quid est et si est non essent questiones, quod est contra Philosophum dicentem quod questiones sunt quattuor, scilicet quia est, quid est, etc. Et etiam istud quod dicunt quod de immediatis non est questio, hoc est falsum, quia de primis principiis est questio et de substantiis separatis. Et ad hoc quod confirmant 'omnis questio est questio medii', non intelligunt Philosophum. Non enim dicitur questio medii quia querit medium, quia solum una questio querit medium, ut questio propter quid, sed sic: quod omnis questio querit medium, quia scito medio absolvitur omnis questio. Cognito enim medio demonstrationis, quod est quid subiecti, statim cognoscitur subiectum; hoc autem cognito cognoscitur si est; hoc autem cognito cognoscitur quod passio inest subiecto et propter quid inheret, ita cognito medio absolvitur omnis questio.

Sed tu dices: quid est medium in demonstratione? Dico quod quod quid est subiecti vel definitio dicens quod quid est subiecti. Et cognito quid est subiecti cognoscitur ipsum si est, ut si est aliqua essentia. Item, cognosci-

---

49 medio] subiecto M

part unknown, and all such can have questions raised about them. For whoever asks also doubts, and everyone who doubts is midway between being ignorant and being a knower. And therefore it is said that in every question it is necessary to presuppose something and to ask something, and so everything about which a question is raised is in part known and in part unknown. It is right then, for the Philosopher to investigate the number of knowables together with the number of things about which questions can be raised.

But it should be understood that some intend to make the number of questions the same as the number of things knowable through demonstration. And their reason is that every question is about things with a middle term that has a cause, for the Philosopher goes on to say that every question is a question about the middle term, and a cause is the same as a middle term. And therefore they hold that every question is about what has a middle term and a cause, but all those that have a middle term are knowable through demonstration, and therefore they maintain that there are questions only of such knowables.

But if this were true then there would be no question of what something is or if something is. This is not known through demonstration but is instead possessed before every demonstration. If, then, a question were only about what is known through demonstration, then the questions what-it-is and if-it-is would not be questions, which is opposed to the Philosopher, who holds that there are four questions, namely whether-it-is-so, what-it-is, and so on. And also, this that they say—that there is no question about immediate things—is false, since there is a question about first principles and about separated substances. And as for this that they affirm—that every question is a question about the middle term—they do not understand the Philosopher correctly. For it is not called a question about the middle term because it seeks out a middle term, for only one question seeks out the middle term, the question because-of-what. Instead, every question seeks a middle term because if the middle term is known then every question is resolved. For if the middle term of demonstration is known, which is the *what* of the subject, then the subject is immediately cognized. But once this is cognized, it is cognized if-it-is, and once that is cognized it is cognized that the attribute is in the subject and because-of-what it inheres. So once the middle term is cognized, every question is resolved.

But you will reply, 'What is the middle term in demonstration?' I answer that the essence of the subject, or the definition indicating the essence of the subject, and once the essence of the subject is cognized the question

tur utrum aliqua passio sibi insit, quia quod quid est subiecti est causa passionis, et cognoscitur quia est. Item, cognito ipso cognoscitur propter quid passio inest subiecto.

Sed tu dices: Philosophus dicit inferius quod omnis questio querit causam; et causa et medium idem; ergo omnis questio querit medium. Dico quod verum est quod omnis questio querit causam, non tamen querit causam sub ratione cause, sed solum questio propter quid. Tamen alie questiones querunt illud quod est causa, ut questio si est querit illud quod est causa; querit enim essentiam rei, ista autem essentia est medium et causa, sed non querit causam sub ratione cause et medii.

Dicendum est ergo alio modo quod Philosophus sumit hic numerum questionum per numerum scibilium communiter.

Sed intelligendum quod quedam sunt scibilia que solum sciuntur per demonstrationem, et talia sunt conclusiones demonstrationis, quedam autem sciuntur, quia cognoscuntur ante omnem demonstrationem, et ita cognoscuntur prima principia. Ideo dicendum quod numerus questionum sumitur a numero scibilium secundo modo, non tamen primo modo, dicendo quod questiones sunt equales numero his: 'quecumque vere cognoscimus', et ideo non exponitur li 'vere' adverbialiter sed coniunctionaliter, unde translatio greca habet 'his que quidem scimus'.[a] Et hoc videtur esse intentio Averrois[b] super istum passum: scibilium enim quedam sciuntur per totam demonstrationem, et sic scitur quia est, ut quia passio inest subiecto; quedam per medium demonstrationis, et sic scitur propter quid, medium enim est causa quare passio inest subiecto; quedam autem per definitionem, que est medium, non tamen in quantum est medium, ut quid est non scitur neque per demonstrationem neque per medium sed per definitionem que est medium, est enim definitio sermo dicens quid est res, et non dicit esse vel non esse, et similiter si est, quia scito quid est, scitur si est. Unde dicit Alpharabius[c] quod questio si est non determinatur per propositionem categoricam sed per quandam conditionalem, ut si cognoscitur quid est, cognoscitur si est, et non loquor de esse in effectu.

---

[a] Cf. Aristoteles latinus, *Analytica posteriora*, trans. Iacobus Veneticus, 2, 1, p. 69: 'Questiones sunt equales numero quecumque vere scimus'.   [b] Averroes, *Expositio magna in Analytica posteriora* (1562, ff. 402D–403A)   [c] Cf. Albertus Magnus, *In Analytica posteriora*, II, I, 1 (ed. Borgnet, p. 156)

---

58 querit¹]   medium *add. sed del.* M   68 ita]   sciuntur *add. sed del.* M

of whether that subject exists is cognized, namely, whether some essence exists. Further, it is cognized whether any attribute inheres in it, since the essence of the subject is the cause of the attribute, and so it is cognized that it is so. Further, once this is cognized, it is cognized because-of-what the attribute is in the subject.

If you reply, 'The Philosopher says below that every question seeks the cause, and the cause and the middle term are the same; therefore every question seeks the middle term', I answer that it is true that every question seeks the cause, but nevertheless it does not seek the cause under the aspect (*ratio*) of cause, but only the question because-of-what. Nevertheless, other questions seek that which is the cause. For instance, the question if-it-is seeks that which is the cause, for it seeks the essence of the thing. This essence is the middle term and the cause. But it does not seek the cause under the aspect of cause and middle term.

We should, then, give a different reply: that the Philosopher here takes the number of questions to match the number of knowables in a broad sense.

It should be understood that some knowables are known only through demonstration, and these are the conclusions of a demonstration. Others are known by being cognized before every demonstration, and this is how first principles are cognized. Therefore it must be held that the number of questions matches the number of knowables in the second way but not in the first way, when he says that the questions are 'equal in number to these which truly we cognize'. (And this 'truly' is not taken adverbially but conjunctively. Hence the Greek translation has 'whatever, indeed, we cognize', and this seems to be the understanding of Averroes in this passage.) Of knowables, some are known through the whole demonstration, and it is known in this way that it is so, that is, that an attribute inheres in the subject. Others are known through the middle term of a demonstration, and this is how the question because-of-what is known. For the middle term is the cause why the attribute inheres in the subject. Still others are known through the definition, which is the middle term, but not inasmuch as it is the middle term. For instance, the essence is not known through demonstration, nor through the middle term, but through the definition, which is the middle term. For a definition is an expression indicating what a thing is, and it does not indicate being or non-being. The same holds for the question of whether it exists, for once the essence is known it is known whether it exists. Whence Alfarabi says

Unde dicit hic Albertus[a] quod questiones sunt equales numero his que scimus, questiones, inquam, non secundum materiam et numerum, quia tales sunt multo plures quam ea que scimus, sicut possumus querere de contingentibus et futuris et aliis, sed secundum speciem, que reducitur in quaternarium numerum. Unde | omnis sciens aut scit quid est aut si est aut quia est aut propter quid est, similiter omnis querens aut querit quid est aut propter quid est aut si est aut quia est.

M 108rb

Ad rationes.

Ad 1.1 Ad primam, nego maiorem. Ulterius cum dicis 'de non ente etc.', dico quod possumus sumere non ens dupliciter: vel non ens simpliciter, et de tali non est scientia nec questio, vel possumus accipere non ens quod non est in effectu, est tamen in suis causis, et de tali non ente est scientia et questio. Et cum dicit Philosophus 'quod non est non convenit scire', verum est, quod nullo modo est.

Ad 1.2 Ad aliam rationem, dico quod de primis principiis est questio. Et cum dicis quod de eis non est scientia, dico quod verum est quod de eis non est scientia proprie dicta, tamen de eis est scientia communiter dicta.

Ad 1.3 Ad tertium, cum dicitur 'que precognoscimus non querimus', dico quod falsum est; contingit enim scire quod triangulus habet tres, hoc autem cognito contingit querere de triangulo si est, postmodum habita cognitione si est bene contingit querere propter quid triangulus habet tres, unde de eodem bene est questio et scientia.

Ad 1.4 Ad ultimam rationem, dico quod verum est quod Philosophus dicit in fine VII *Metaphysice* quod in simplicibus non est questio, cum non habeant partem et partem, et in omni questione oportet habere partem et partem, quia aliquid oportet querere [et aliquid oportet querere] et aliquid supponere. Ideo in simplicibus ut sic non est questio, unde considerando ipsas secundum suas naturas, sic de eis non est questio. Possunt

---

[a] Albertus Magnus, *In Analytica posteriora*, II, I, 1 (ed. Borgnet, p. 156)

---

86 quia] sed M   88 futuris] fort[tis] M   96 est[2]] ens M   102 que] quem M   105 propter quid] utrum M   109 oportet] querere *add. sed del.* M

that the question of if-it-is is not settled through a categorical proposition, but through a kind of conditional, so that if what-it-is is cognized then if-it-is is cognized. And here I am not speaking of being in actual effect.

Hence Albert says here that questions are equal in number to the things we know—questions not according to matter and number, for there are many more such than what we know, given that we can ask about contingent and future things among others, but according to species, which reduce to four. Hence every knower either knows what-it-is or if-it-is or that-it-is-so or because-of-what-it-is. In the same way, everyone asking a question either asks what-it-is or because-of-what-it-is or if-it-is or whether-it-is-so.

In response to the arguments.

Ad 1.1 In response to the first I deny the major premise. Next, when you say, 'concerning a non-being' etc., I reply that we can take non-being in two ways, either as a non-being without qualification, and concerning such there is neither knowledge nor any question, or we can take non-being as what does not exist in actual effect but instead exists in its causes, and concerning such a non-being there is knowledge as well as a question. And when the Philosopher says that knowing does not apply to what is not, this is true of what does not exist in any way.

Ad 1.2 In response to the second argument I reply that there is a question concerning first principles, and when you say that there is no knowledge about these, I reply that it is true that there is no knowledge about these strictly speaking, but there is knowledge about them in a broad sense.

Ad 1.3 In response to the third argument, when it is claimed that what we cognize in advance we do not ask about, I reply that this is false. For one can know that a triangle has three etc., and knowing this, one can ask about the triangle whether it exists. Then, possessing the cognition of whether it exists, it plainly becomes possible to ask because-of-what a triangle has three etc. Hence there plainly is both question and knowledge about the same thing.

Ad 1.4 To the last argument I reply that it is true that the Philosopher says in the end of *Metaphysics* VII that there is no question in the case of simple things, since they do not have part beyond part, and in every question there has to be part beyond part, since one has to ask one thing and to suppose another. Therefore in the case of simple things as such there is no question. Hence, there is no question about them when considering

tamen considerari ut sunt cause aliquorum effectuum, ut motuum, ⟨et⟩ sic possunt esse secundum aliquid cognite, ut quantum ad earum operationes, et secundum aliquid ignote, ut secundum suas substantias, et ideo de eis ut sic potest esse scientia et questio.

⟨Questio 44⟩

Consequenter queritur utrum si est sit questio ponens in numerum.

1.1 Et arguitur quod sic, quia illa questio ⟨que⟩ querit de inherentia unius ad alterum est questio ponens in numerum; sed questio si est est questio querens inherentiam unius ad alterum; ergo etc. Maior patet; ubi enim est inherentia unius ad alterum est comparatio duorum ad invicem, et per consequens ibi est numerus.

1.2 Item, illa est questio ponens in numerum in qua illud quod queritur et illud de quo queritur sunt diversa secundum essentiam; sed in questione si est illud quod queritur et de quo queritur sunt diversa secundum essentiam, scilicet esse et cuius est esse, cuius probatio est quia distincte comprehendit⟨ur⟩ unum preter alterum; ergo etc.

1.3 Item, questio si est aut est questio subiecti aut passionis: non subiecti, quia de ipso non est querere si est, sed illud oportet precognoscere, ergo est questio passionis; sed questio passionis non est simplex, immo concernit subiectum; ergo etc.

2 Oppositum vult Philosophus[a] dicens quod questio si est est questio simplex non ponens in numerum.

3 Intelligendum est hic quod questio si est est questio, sed non est questio ponens in numerum. Et quod sit questio declaratur sic: in omne questione oportet aliquid esse suppositum et aliquid ignoratum; omnis enim qui querit dubitat et dubitans medius est inter scientem et ignorantem, aliquid novit de re et aliquid ignorat; in omni ergo questione est aliquid scitum

---

[a] Aristoteles, *AnPo*, II, 1, 89 b 32–33

18 questio²] simplex *add. sed del.* M

them according to their natures, but they can be considered as they are causes of various effects, such as motions. Thus they can be cognized according to something, for instance their operations, while according to something else, for instance their substances, they are unknown. And therefore there can be knowledge of simple things in this way, and there can be a question as well.

Question 44.

Next it is asked whether the question of if-it-is is a question implying a numerical difference.

1.1 It is argued that it is, since a question which asks about the inherence of one to another is a question implying a numerical difference. But the question of if-it-is is a question asking about the inherence of one to another. Therefore etc. The major premise is obvious, for where there is an inherence of one to another there is a relation of two to one another, and consequently there is a number difference.

1.2 Furthermore, that is a question implying a numerical difference in which what is asked and that about which it is asked are different in essence. But in the question of if-it-is, what is asked and that about which it is asked, namely being and that of which the being belongs to, are different in essence. The proof is that one is distinctly comprehended without the other. Therefore, etc.

1.3 Furthermore, the question of if-it-is is a question either about the subject or about the attribute. It is not about the subject, since one does not ask if-it-is concerning this; rather, it is necessary to cognize that in advance. Therefore it is a question about the attribute. But a question about the attribute is not simple, but concerns the subject. Therefore, etc.

2 On the other hand, the Philosopher says that the question of if-it-is is a question about a simple and not one implying a numerical difference.

3 It should be understood that the question of if-it-is is a question about a simple, and not a question implying a numerical difference. And that it is a question is explained as follows: in every question it is necessary that something be assumed and something be unknown, for everyone who raises a question doubts, and one who doubts is in between one

et aliquid ignoratum. Sed in questione si est est aliquid suppositum, supponitur enim quod illud quod significatur per nomen habet esse simpliciter, et ignoratur utrum sit aliqua essentia vel natura. Et propter hoc questio si est est questio.

Sed estne questio ponens in numerum? Hic est intelligendum quod omne queribile est scibile, et omne scibile enuntiabile, quia omne scibile est verum aut falsum, et ideo omnis questio reducitur ad aliquem modum enuntiandi. Sed duplex est modus enuntiandi: unus in quo hoc verbum 'est' predicatur secundum adiacens, alius in quo predicatur hoc verbum 'est' tertium adiacens. Et omnis ⟨questio⟩ que refertur ad modum enuntiandi in quo predicatur hoc verbum 'est' tertium adiacens est questio ponens in numerum, dummodo illud quod queritur et de quo queritur sint diversa secundum essentiam.

Sed dicet aliquis: per te idem est secundum subiectum questio et enuntiatio; sed si questio et enuntiatio sint idem secundum subiectum, cum omnis enuntiatio sit composita, ergo omnis questio est composita, quod falsum est secundum Philosophum; falsum est ergo dicere quod omnis questio reducitur ad enuntiationem.

Dico quod in omni questione est ratio compositi, quoniam in omni questione est aliquid scitum et aliquid ignoratum, sed propter hoc non oportet quod omnis questio sit questio ponens in numerum. Probatio, quia ubicumque est questio ponens in numerum, ibi illud quod queritur est diversum secundum essentiam ab illo de quo queritur, ut utrum homo sit risibilis; sed esse simpliciter dictum non est diversum ab essentia uniuscuiusque, entitas enim uniuscuiusque includitur in essentia uniuscuiusque; et ideo, cum talis entitas absoluta queratur per questionem si est, questio si est non dicitur questio ponens in numerum. Et quod entitas sit in essentia uniuscuiusque patet per Commentatorem[a] IV *Metaphysice*. Cum enim homo est ens, aut ergo per essentiam suam aut per additum, si per | essentiam suam, habetur propositum; si per additum, oportet illud additum esse ens, quia per non ens non est aliquid ens; tunc queram de illo, aut est ens per se aut per aliquid additum, et sic in infinitum. Et iterum si res est ens per aliquid additum, illud additum vel erit de genere substantie vel de genere accidentis. Non de genere substantie, quia ibi inveniuntur tres, scilicet materia, forma aut compositum, sed nullum istorum est additum

M 108va

---

[a] Averroes, *Commentarium magnum in Metaphysicam Aristotelis*, IV, 2, comm. 3 (1562, ff. 67CB, GH).

---

25 simpliciter] s^m M   36 questio] quid M   37 sed] quare M   50 cum] dicam M   51 essentiam] entitatem M   51–52 essentiam] entitatem M   57 compositum] compositio *sed* io *exp.* M

who knows and one who does not. He recognizes something about the thing and is ignorant of something else. So in every question something is known and something is unknown. But in the question of if-it-is there is something assumed. For it is assumed that what is signified by the name has being without qualification and it is not known whether there is an essence or nature. And so the question of if-it-is is a question.

But is it a question implying a numerical difference? Here it must be understood that everything that can be asked about is knowable and every knowable can be asserted—for every knowable is true or false—and so every question goes back to some mode of assertion. But there are two modes of asserting, one in which the word 'is' is predicated as a second adjacent, the other in which the word 'is' is predicated as a third adjacent. And every question that is referred to in that way of asserting in which the word 'is' is predicated as a third adjacent is a question implying a numerical difference, as long as what is asked and that about which it is asked differ in essence.

But someone will say, 'According to your view a question and an assertion are the same in their subject. But if a question and an assertion are the same in subject, since every assertion is composite, therefore every question is composite, which is false according to the Philosopher. It is false, then, to say that every question leads back to an assertion'.

I reply that every question has the account of composition, because in every question there is something known and something unknown, but it is not necessary on this account that every question be a question implying a numerical difference. Proof: Since wherever there is a question implying a numerical difference, that which is asked is different in essence from that about which it is asked, for instance, whether a human being is able to laugh, although it is not said without qualification that each is different from the essence of the other, for the being of each is included in the essence of the other. And therefore, since such an absolute being is asked about through the question of if-it-is, the question of if-it-is is not called a question implying a numerical difference. And that the being is in the essence of the other is clear from the Commentator on *Metaphysics* IV. For since a human being is a being, he is so either through his essence or through something added. If through his essence, we have what was proposed. If through something added, it is necessary that the added something be a being, for through a non-being there is no being. Then I will ask about that being, whether it is a being *per se* or through something

substantie rei. Nec de genere accidentis, quia si sic, idem sibi ipsi erit causa essendi: quia esse cuiuslibet accidentis est causatum a substantia, si tunc dicas quod substantia est ens per aliquod accidens, substantia erit causa illius accidentis; sed esse illius accidentis est causatum a substantia, ergo substantia causat illud esse quo est; ergo idem erit causa essendi sui ipsius, quod est impossibile.

Manifestum est ergo quod questio si est non est questio ponens in numerum, et loquor de esse simpliciter, quia de esse in effectu est alia ratio. Si tu queras de esse in effectu, verum est dicere quod est additum rebus causatis pro tanto, quia nullum causatum est ipsum suum esse, verumtamen esse in effectu non est res addita rebus causatis. Quantum ergo ad hoc quod esse in effectu est quid additum, questio si est est questio ponens in numerum, quantum tamen ad hoc quod esse in effectu non est res addita rebus causatis, questio si est in effectu non est questio ponens in numerum.

Sed tu queres: qualiter ista duo stant quod esse in effectu sit additum et non sit res addita? Similitudo fundata supra albedinem non est res aliqua addita; aliquid enim non est compositi[v]us quamvis sit album et simile quam si sit album solum, et tamen albedo non est similitudo. Eodem modo, quamvis esse in effectu non sit res addita ei cuius est, non tamen esse in effectu est illud cuius est.

Ad rationes.

Ad 1.1 Ad primam, cum arguitur 'questio in qua est inherentia etc.', dico quod falsum est, immo questio in qua est inherentia unius ad alterum – et sunt diversa secundum essentiam illud quod queritur et de quo queritur – est questio ponens in numerum. Sed quando illud quod queritur et de quo queritur non sunt diversa secundum essentiam, talis est questio simplex, et talis est questio si est.

Ad 1.2 Ad aliam, nego minorem; probatum enim est quod entitas uniuscuiusque est in esse⟨ntia⟩ ipsius. Sed tu dicis: nonne possum intelligere

---

70 esse in effectu] questio si est M   71 causatis] -tis *dub.* M   74 similitudo] similimodo *ante corr.* M   85 simplex] simpliciter M

added, and so on indefinitely. And again, if a thing is a being through something added then that added thing will either be of the genus of substance or the genus of accident. It is not from the genus of substance, for three things are found there: matter, form, and the composite. But none of these is something added to the substance of a thing. Nor is it of the genus of accident, for if it were then something would be the cause of its own being, since the being of every accident is caused by a substance. If, then, you reply that a substance is a being through some accident, the substance will be the cause of that accident. But the being of that accident is caused by the substance, therefore something will be the cause of its own being, which is impossible.

So it is obvious then that the question of if-it-is is not a question implying a numerical difference. And I speak of being without qualification, for there is another argument concerning being in actual effect. If you ask about being in actual effect, it is truly said that it is something added to things inasmuch as they are caused, since nothing caused is its own being. But nonetheless being in actual effect is not a thing added to caused things. Insofar, then, as being in actual effect is something added, the question of if-it-is is a question implying a numerical difference; but insofar as being in actual effect is not a thing added to caused things, the question of if-it-is in actual effect is not a question implying a numerical difference.

But you will ask: how do these two things stand, i.e., that being in actual effect is added and yet that it is not an added thing? I reply that a similarity founded upon whiteness is not some added thing, for something is not more composite if it is both white and similar than if it is only white. And yet whiteness is not similarity. In the same way, even though being in actual effect is not a thing added to that to which it belongs, yet being in actual effect is not that to which it belongs.

In response to the arguments.

Ad 1.1 In reply to the first, when it is argued 'a question in which there is inherence etc.', I reply that this is false. Instead, a question in which there is inherence of one to another, and in which what is asked and that which it is asked about are different in essence, is a question implying a numerical difference. But when what is asked and that about which it is asked are not different in essence, then that sort of question is simple. And this is how it is for the question of if-it-is.

Ad 1.2 In response to the next, I deny the minor premise. For it was proved that the being of each thing lies in its essence. But you will reply,

rem non intelligendo esse absolutum? ⟨Dico quod non preter esse absolutum⟩, preter tamen esse in effectu potest intelligi, quare questio si est secundum quod querit de esse in effectu uno modo est questio ponens in numerum.

Ad 1.3 Ad rationem, dico quod questio si est est questio subiecti, passionis et medii. Unde per questionem si est tu potes [intelligere] querere si est subiectum, et si est medium, et si est passio. Et ista se habent secundum ordinem, quia esse passionis presupponit esse subiecti, esse subiecti presupponit esse medii, et esse medii presupponit esse simpliciter. Et de quolibet istorum est aliquid querere, quia de quolibet est aliquid suppositum et aliquid ignoratum. Questio enim si est passionis presupponit esse subiecti et querit esse passionis, questio si est subiecti presupponit esse medii et querit esse subiecti, questio si est medii presupponit esse simpliciter et querit esse medii. Et esse simpliciter non contingit querere, quia esse simpliciter est illud quod primo occurrit intellectui. Verumtamen de esse in effectu vel de esse in anima bene est questio, de esse tamen simpliciter non est questio, sed hoc presupponitur in unoquoque.

Et cum dicitur 'si est est precognitio etc.', dico quod illud quod primo est questio potest esse precognitio; primo enim queritur si est et post potest esse precognitio respectu alicuius alterius.

Et cum dicitur ulterius 'esse passionis non est aliquid simplex sed concretum subiecto', dico quod esse passionis est duplex, scilicet essentie et esse existentie. Esse essentie bene est absolutum ad esse subiecti, immo essentia passionis non est essentia subiecti; que enim faciunt intellectus distinctos et separatos faciunt essentias distinctas. Propter hoc igitur est absolutum quia ad essentiam suam non pertinet essentia subiecti. Et ideo, quia passio quantum ad illud esse non dependet, ideo de passione illo modo potest esse questio si est ponens in numerum. Questio tamen si est que querit de esse existentie ipsius passionis, cum sit additum essentie, est questio ponens in numerum.

---

89 quare] quia M   90 secundum quod] *s. l.* M   ‖ esse] essentia *ante corr.* M   97 quia] et M   ‖ aliquid²] ignoratum *add. sed del.* M   100 presupponit] si *add. sed exp.* M

'Can I not understand the thing while not understanding absolute being?' I say that it cannot be understood outside absolute being, but it can be understood outside being in actual effect. Therefore the question of if-it-is, inasmuch as it asks about being in actual effect in one way, is a question implying a numerical difference.

Ad 1.3 In response to the ⟨last⟩ argument, I maintain that the question of if-it-is is a question about the subject of the attribute and the middle term. Hence through the question of if-it-is you can ask whether the subject exists, whether the middle term exists, and whether the attribute exists. And these are related in a certain order, for the being of the attribute presupposes the being of the subject, the being of the subject presupposes the being of the middle term, and the being of the middle term presupposes being without qualification. And about each of these there is something to ask, since for each there is something assumed and something unknown. For the question of if-it-is, asked of the attribute, presupposes the being of the subject and asks about the being of the attribute. The question of if-it-is, asked of the subject, presupposes the being of the middle term and asks about the being of the subject. The question of if-it-is, asked about the middle term, presupposes being without qualification and asks about the being of the middle term. And being without qualification cannot be asked about, for it is that which first occurs to the intellect. But there certainly can be a question about being in actual effect or being in the soul. Concerning being without qualification there is no question, but this is presupposed in every question.

And when it is said that 'if-it-is is a precognition' etc., I reply that what is first a question can become an advance cognition. For first it is asked whether the thing exists, and afterwards it can become an advance cognition with respect to something else.

And when it is said next that the being of an attribute is not something simple, but contracted to a subject, I answer that the being of an attribute can be taken in two ways, namely as the being of the essence and the being of existence. The being of the essence is surely without qualification with respect to the being of the subject, but the essence of the attribute is not the essence of the subject, for what produces distinct and separate concepts produces distinct essences. Because of this, then, it is without qualification, for the essence of the subject does not pertain to its essence. Therefore, since an attribute is not dependent with respect to that being, there can be, about an attribute in that way, a question of if-it-is that implies a numerical difference. But the question of if-it-is that asks about the being of the attribute's existence, since it is added to the essence, is a question implying a numerical difference.

⟨Questio 45⟩

Consequenter queratur de questione quid est, et queratur utrum questio quid est sit questio.

1.1 Et arguitur quod non, quia quale est et quantum est non ⟨est⟩ questio, [questio] ergo nec quid est. Consequentia patet de se. Antecedens etiam patet, quia quale est et quantum est non numerantur inter questiones.

1.2 Item, quod est precognitio non est questio; sed quid est est precognitio; ergo quid est non est questio.

2 Oppositum patet per Philosophum.[a]

3 Et est intelligendum quod questio quid est est questio, quia de omni eo contingit querere de quo contingit dubitare; sed de quidditate rei contingit dubitare. Nam de omni eo contingit dubitare cuius ratio est posterior ratione entis; ens enim est illud quod primo est cognitum intellectui nostro, et manifestum est quod de omni eo contingit dubitare quod est posterius primo | intellecto, ideo de omni eo contingit dubitare quod est posterius ente. Sed quid est est posterius ipso ente, propter hoc dixit Avicenna[b] quod res certificantur per ens, quia si queratur quid est res, respondetur quod est ens. Si ergo quid est est posterius ente, de eo contingit querere. Et ex hoc sequuntur duo: unum est quod questio quid est est distincta a questione si est, et aliud est quod questio quid est presupponit questionem si est. Et utriusque declaratio est, quia questio si est querit de entitate rei absoluta et questio quid est querit de quidditate rei determinata; cum enim quero quid est, quero in quo genere est res; et quia quidditas rei talis dependet ab esse simpliciter, ideo questio quid est presupponit questionem si est. Et hoc dicit Philosophus[c] postea, quod querentem quid est necesse est cognoscere si est.

M 108vb

Sed ulterius queret aliquis: dato quod questio quid est sit questio, estne questio ponens in numerum? Dico quod non, quia illa est questio ponens

---

[a] Aristoteles, *AnPo*, II, 1, 89 b 24–25 [b] Avicenna latinus, *De Philosophia prima vel scientia divina*, I, 5 (ed. Van Riet, p. 35) [c] Aristoteles, *AnPo*, II, 2, 90 a 8–10

---

14 intellecto]   intento M   22 quidditas]   entitas M   24 querentem]   cognoscentem M

Question 45.

Next let us ask about the question of what-it-is: whether the question of what-it-is is a question.

1.1 It is argued that it is not, since what sort it is and how much it is are not questions; therefore what-it-is is not a question. The inference is obviously correct, and the antecedent is also obvious since what sort it is and how much it is are not found among the questions.

1.2 Furthermore, whatever is an advance cognition is not a question, but what-it-is is an advance cognition; therefore what-it-is is not a question.

2 The opposite view is clear from the Philosopher.

3 And it should be understood that the question of what-it-is is a question, for it is possible to ask a question about everything about which it is possible to have a doubt, but it is possible to have a doubt concerning the quiddity of a thing, for it is possible to have a doubt about everything whose account is posterior to the account of being. For being is that which is first cognized by our intellect, and it is obvious that it is possible to have a doubt about everything which is posterior to what is understood first. Therefore, doubt can arise concerning everything that is posterior to being. But what something is is posterior to being. For this reason Avicenna said that things are made definite through being, for if it is asked what a thing is it will be responded that it is a being. If, then, what something is is posterior to being, then it is possible to raise a question about it. And from this two things follow. One is that the question what-it-is is distinct from the question of if-it-is. And the other is that the question of what-it-is presupposes the question of if-it-is. And the explanation for both of these is that the question of if-it-is asks about the absolute being of a thing, and the question of what-it-is asks about the determinate quiddity of the thing. For when I ask what-it-is I ask in what genus the thing is, and since the quiddity of such a thing depends on being without qualification, therefore the question of what-it-is presupposes the question of if-it-is. The Philosopher goes on to say this: that it is necessary for one who asks what-it-is to cognize if-it-is.

But someone will further ask, 'Given that the question of what-it-is is a question, is it a question implying a numerical difference?' I reply that

in numerum in qua quod queritur et de quo queritur habent essentias distinctas et separatas; sed in questione quid est illud quod queritur et illud de quo queritur non habent essentias distinctas, quia illud quod queritur est definitio rei et illud de quo queritur est entitas rei, nunc autem definitio et entitas non sunt essentie distincte.

Sed aliquis dicet: principium ponit in numerum cum principiato; sed quod quid est est principium rei, unde Philosophus[a] in VII dicit quod quod quid est est principium illius cuius est quod quid est; si ergo hoc queratur, questio quid ⟨est⟩ est questio ponens in numerum.

Item, questio quid est et propter quid non differunt; sed questio propter quid est questio ponens in numerum; ergo etc.

⟨Ad hoc⟩ quod dicitur primo, 'principium etc.', ⟨dico⟩ quod principium ea ratione qua principium, et causa ea ratione qua causa, ponit in numerum cum principiato. Et cum dicitur 'quod quid est est principium', dico quod in quod quid est importantur duo, scilicet materia et forma. Et illud quod principaliter importatur forma est, et ideo partes definitionis forme sunt, quia ratione materie non est magis homo [est] homo quam asinus. Unde considerando ad formam dico quod quod quid est est principium, ⟨et⟩ hoc modo accipiendo habet rationem principii et ponit in numerum cum eo; sed accipiendo quod quid est non solum ⟨ut⟩ formam, sed formam et materiam, non ponit in numerum, nec est principium eius, sed omnino idem cum eo cuius est quod quid est, nec prius nec posterius ipso.

Et cum dicitur 'quid et propter quid idem sunt', dico quod idem sunt secundum subiectum, differunt tamen secundum rationem. Nam quid subiecti comparatum ad subiectum absolute habet rationem quid, comparatum autem ad subiectum ut est causa inherentie predicati ad subiectum meretur hoc nomen 'propter quid', et ideo idem est quod quid est subiecti et dicens causam inherentie passionis ad subiectum. Hinc est quod questio propter quid dicitur questio ponens in numerum, quia includit inherentiam passionis ad subiectum, que differunt secundum essentiam; passio enim et subiectum differunt numeraliter, habent enim essentias numeratas, esse enim subiecti non est essentia passionis nec e converso. Sed questio quid est non includit talem inherentiam duorum quorum unum

---

[a] Aristoteles, *Met.*, VII, 7–9

it is not, for a question implying a numerical difference is one in which what is asked and that about which it is asked have distinct and separate essences, but in the case of the question of what-it-is, what is asked and that about which it is asked do not have distinct essences, for what is sought is the definition of the thing and that about which it is asked is the being of the thing. Now the definition and the being are not distinct essences.

But someone will reply that the principle implies a numerical difference with that of which it is a principle. But the essence of a thing is its principle, hence the Philosopher in *Metaphysics* IV says that the essence of a thing is the principle of that of which it is the essence. If, then, this is asked, then the question of what-it-is is a question implying a numerical difference.

Further, the questions of what-it-is and because-of-what do not differ, but the question because-of-what is a question implying a numerical difference; therefore etc.

In reply to what is said first, 'The principle etc.', I say that the principle considered as principle and the cause considered as cause imply a numerical difference with that of which it is a principle. And when it is said that the essence is a principle, I reply that there are two things conveyed in an essence, namely matter and form. And what is principally conveyed is form. Therefore the parts of a definition are forms, since on account of matter a human being is no more a human being than he is a donkey. Hence, considering the form, I reply that an essence is a principle. In this way of taking it, it has the account of a principle and implies a numerical difference with it. But when one takes an essence not only as form, but as form and matter, it does not imply a numerical difference, nor is it a principle of it but rather entirely identical with that of which it is an essence, nor is it prior to or posterior to it.

And when it is said that the questions of what-it-is and because-of-what are the same, I reply that they are the same according to subject, but differ in account. For the quiddity of the subject related to the subject without qualification has the account of its quiddity, but related to the subject as it is the cause of inherence of the predicate to the subject it merits the name of that because-of-which. And therefore it is the same to speak of the essence of a subject and to speak of the cause of the inherence of an attribute to a subject. And thus the question because-of-what is called a question implying a numerical difference, since it includes the inherence of the attribute to the subject, which differs from it in essence. For attribute and subject differ in number, for their essences

differat ab alio secundum essentiam, ideo est questio simplex. Unde dicit Albertus[a] quod illa questio cuius ultimum est scire aliquid simplex est questio simplex, et illa questio cuius ultimum est scire aliquod complexum est questio ponens in numerum. Nunc autem questio quid est et questio si est intendunt scire aliquod incomplexum, ut essentiam rei, questio propter quid est et quia est intendunt scire inherentiam unius ad alterum et causam inherentie, ideo dicuntur questiones ponentes in numerum.

Item, licet questio quid est et questio si est in hoc conveniant quod utraque est questio simplex, tamen differunt, quia questio quid est presupponit questionem si est, et iterum quid est se habet ad si est sicut definitio ad definitum; definitum autem notius est nobis quam definitio per Aristotelem I *Physicorum*,[b] quia quanto aliquid est magis confusum tanto magis notum; et ideo querentem quid est necesse est cognoscere si est.

Ad rationes.

Ad 1.1 Ad primam, dico quod quale est et quantum est sunt questiones. Et tu dices: ad quam questionem reducuntur? Dico quod ad questionem quia est, quia illa est questio quia in qua queritur aliquid de aliquo quod est in essentia distinctum ab eo; sed cum queritur quale est et quantum est illud quod queritur est distinctum in esse ab eodem de quo queritur; ideo quando queritur quale est aut quantum est est questio quia.

Ad 1.2 Ad aliam rationem, dico quod illud quod est precognitio non est questio eodem modo et respectu eiusdem, tamen illud quod est questio potest esse postea precognitio respectu alicuius alterius, sicut quid est subiectum est primo questio et postea est precognitio ad probandum passionem de subiecto, et ideo potest esse questio et precognitio.

---

[a] Albertus Magnus, *In Analytica posteriora*, II, I, 2 (ed. Borgnet, p. 161) [b] Aristoteles, *Phys.*, I, 1, 184 b 10–12

---

61 simplex] simpliciter M   63 aliquod] aliquid *ante corr.* M   66 est[1]] *s. l.* M   73 querentem] cognoscentem M   81 precognitio] precognitionem *ante corr.* M

are distinctly numbered. For the being of the subject is not the essence of the attribute nor conversely. But the question of what-it-is does not include such an inherence of two things, one of which differs from the other in essence. Therefore it is a simple question. Hence Albert says that that question whose aim is to know something simple is a simple question, and the question whose aim is to know something complex is a question implying a numerical difference. Now the question of what-it-is and the question of if-it-is aim to know something noncomplex, for example, a thing's essence. The questions because-of-what and whether-it-is-so aim at knowing the inherence of one thing to another and the cause of that inherence. Therefore they are called questions that imply a numerical difference.

Furthermore, although the questions of what-it-is and of if-it-is agree in this that both are simple questions, still they differ since the question of what-it-is presupposes the question of if-it-is, and further what-it-is stands to the question of if-it-is as the definition stands to what is defined. But what is defined is better known to us than the definition, according to Aristotle in *Physics* I, for to the extent that something is more confused it is better known. And therefore someone asking what-it-is necessarily cognizes if-it-is.

In response to the arguments.

Ad 1.1 To the first I reply that 'what sort is it' and 'how much is it' are questions. You will then ask, 'To what question do they reduce?' I reply that they reduce to the question of whether-it-is-so, since that is the question in which something is asked about something which is in essence distinct from it.is But when it is asked what sort is it and how much is it, that which is asked is distinct in being from that about which it is asked. Therefore, when it is asked what sort is it and how much is it, this is the question of whether-it-is-so.

Ad 1.2 In response to the other, I reply that that which is an advance cognition is not a question in the same way and with respect to the same thing, but that which is a question can afterward become an advance cognition with respect to something else, as what the subject is is first a question and then afterwards an advance cognition for proving the attribute of the subject. And therefore it can be both a question and an advance cognition.

⟨Questio 46⟩

Consequenter queritur utrum omnis questio sit questio medii.

1.1 Et arguitur quod non, quia si omnis questio esset questio medii, et omnis questio medii est questio propter quid, ergo omnis questio est questio propter quid. Consequens est falsum, ergo et antecedens. Quod autem omnis questio medii et questio propter quid sint idem, probatio, quia idem [et] est medium et propter | quid, ergo questio medii est questio propter quid. Resumatur tunc ratio.

M 109ra

1.2 Item, omnis questio querit aliquid de aliquo et ita questio est alicuius compositi; sed medium est quid simplex, ut quod quid est subiecti; ergo etc.

2.1 Oppositum patet per Philosophum.[a]

2.2 Et arguitur ratione, quia omnis illa questio est questio medii que potest terminari per medium; sed omnis questio determinatur per medium, quia cognito medio cessat omnis questio.

3 Dicendum est ad hoc quod omnis questio est questio medii, quia omnis questio querit causam; et causa et medium idem; ergo etc. Maior patet inducendo in omnibus, quia questio quid est querit causam formalem, querit enim quidditatem; questio propter quid querit causam inherentie predicati ad subiectum, et ita causam; et questio si est querit causam, quia questio quid est et si est idem querunt substantialiter, sed quod questio quid est querit magis determinate, illud questio si est querit minus determinate; item, et questio quia est querit causam, sicut enim questio quid est querit quid est causa, ita questio quia querit si est causa.

Et rationabiliter hoc contingit secundum Albertum,[b] quia omnis volens devenire ad aliquem finem intendit medium per quod potest devenire in finem illum. Et ideo cum in omni questione queratur finis aliquis,

[a] Aristoteles, *AnPo*, II, 2, 90 a 5–6 [b] Albertus Magnus, *In Analytica posteriora*, II, I, 3 (ed. Borgnet, p. 162)

---

25 quod] quem M

Question 46.

Next it is asked whether every question is a question about a middle term.

1.1 It is argued that not every question is, for if every question were a question about a middle term, and every question about a middle term is the question because-of-what, then every question is the question because-of-what. The consequent is false, therefore so is the antecedent. The proof that every question about the middle term and every question because-of-what are the same is that the middle term and that because-of-which are the same, and therefore a question about the middle term is the question because-of-what. The argument goes on from there.

1.2 Furthermore, every question asks something about something and so a question is of something composite, but the middle term is something simple, the essence of the subject; therefore etc.

2.1 The opposite view is clear from the Philosopher.

2.2 It is also argued for by reason, since every question is a question about a middle term which can be determined through the middle term, but *every* question is determined through the middle term, for once the middle is cognized every question ceases.

3 It should be replied to this that every question is a question about a middle term, for every question seeks the cause and the cause is the same as the middle term; therefore etc. The major premise is obvious by induction on the sorts of questions, for the question of what-it-is seeks the formal cause, since it seeks the quiddity; the question of because-of-what seeks the cause of the inherence of the predicate in the subject and so seeks the cause; the question of if-it-is seeks the cause since the question of what-it-is and the question of if-it-is seek the same thing in substance. But what the question of what-it-is seeks more determinately, the question of if-it-is seeks less determinately. Furthermore, the question of whether-it-is-so seeks the cause. For just as the question of what-it-is seeks what the cause is, so the question of whether-it-is-so seeks whether there is a cause.

All of this is reasonable according to Albert, for everyone wishing to arrive at some end seeks the means through which it can be attained. And therefore, since in every question some end is sought, it is obvious

manifestum est quod in omni questione oportet ire per medium, et ita omnis questio erit questio medii. Sed secundum Albertum[a] medium potest queri duobus modis, quia aut queritur illud quod est medium aut medium sub ratione medii. Questio quid est querit quid est medium, et si est similiter, sed non sub ratione medii, sed questio propter quid est et quia est querunt medium sub ratione medii, et ideo questio propter quid est maxime questio medii, cum querat medium sub ratione medii et magis querat medium quam questio quia. Non est ergo intelligendum quod omnis questio sit questio medii quia querit medium sub ratione medii, sed aliqua querit illud quod est medium absolute, ut visum est.

Et item alia ratio est, quia omnis questio est illius per quod determinatur; sed omnis questio determinatur per medium, quia medium est quod quid est subiecti, et cognito quod quid est subiecti cognoscitur quid est subiectum, et hoc cognito cognoscitur si est, quia questio quid est presupponit si est.

Item, cognito quid est subiecti cognoscitur utrum aliquid sibi insit, et cognito hoc cognoscitur propter quid ei inest, quare cognito medio absolvitur omnis questio.

Ad rationes in oppositum patet solutio.

Ad 1.1 Ad primam, dico quod non omnis questio medii est questio propter quid, sed solum illa que querit medium sub ratione medii; alie enim questiones querunt de medio absolute, ut visum est.

Ad 1.2 Ad aliam, dico quod omnis questio est questio compositi pro tanto, quia non fit questio nisi queratur aliquid de aliquo, tamen non est ita questio compositi quod per compositum terminaretur, immo questio determinatur et absolvitur per simplex. Ut si queratur quid est homo, respondetur quod est animal rationale, et ista significant quid, quod consistit in indivisibili, ut substantiam rei que in indivisibili consistit, et similiter questio si est. Et ita per aliquod simplex absolvitur et determina[n]tur, et ideo sunt questiones simplices.

[a] Albertus Magnus, *In Analytica posteriora*, II, I, 2 (ed. Borgnet, p. 159)

32–33 propter quid] medii M   37 illius] illud M

that in every question it is necessary to proceed through some middle term. And so every question will be a question about the middle term. But according to Albert the middle term can be sought in two ways, for either that is sought which is the middle term, or the middle term is sought under the account of the middle term. The question of what-it-is seeks what the middle term is, and so does the question of if-it-is, but not under the account of the middle term. But the questions of because-of-what and whether-it-is-so seek the middle term under the account of the middle term, and therefore the question because-of-what is most of all a question about the middle term, since it seeks it under the account of the middle term and when it seeks the middle term more than does the question of whether-it-is-so. It is not therefore to be understood that every question is a question about a middle term because it seeks the middle term under the account of the middle term. Instead, some seek that which is the middle term without qualification, as we have seen.

There is also another argument, for every question concerns that through which it is determined. But every question is determined through the middle term. This is because the middle term is the essence of the subject, and once the essence of the subject is cognized it is cognized what the subject is, and once this is cognized it is cognized whether the subject is, for the question of what-it-is presupposes the question of if-it-is.

Furthermore, once it is cognized what the subject is, it is cognized whether something inheres in it and once this is cognized it is cognized because-of-what something inheres in it. Therefore once the middle term is cognized, every question is resolved.

The solution to the opposite arguments is clear.

Ad 1.1 To the first I reply that not every question about the middle term is a question because-of-what-it-is, but only that which seeks the middle term under the account of the middle term, for the other questions ask about the middle term absolutely, as we have seen.

Ad 1.2 In response to the other I reply that every question is a question about a composite inasmuch as no question arises unless something is asked about something. A question about a composite, however, is not such as would be determined through a composite. Instead, the question is determined and resolved through something simple. For instance, if it is asked what a human being is, the response will be that it is a rational animal. And this signifies something that consists in an indivisible, for instance the substance of a thing, which consists in an indivisible. And in the same way the question of if-it-is is also resolved and determined through something simple. And therefore these are simple questions.

⟨Questio 47⟩

Consequenter queritur de medio demonstrationis, et primo utrum definitio passionis sit medium in demonstratione, secundo utrum definitio subiecti.

1.1 Et quod definitio passionis apparet, quia medium in demonstratione, sicut dicit Philosophus,[a] est ratio primi termini; sed primus terminus in demonstratione est maior extremitas et passio; quare etc.

1.2 Item, medium in demonstratione debet esse medium secundum naturam et non solum ⟨secundum rationem⟩, ergo illud erit medium quod est prius minore extremitate et posterius maiore extremitate; sed tale est definitio passionis, ipsa enim posterior est maiore extremitate, scilicet ipsa passione, et est prior minore extremitate, ut subiecto; ergo etc.

1.3 Item, medium in demonstratione debet esse eiusdem nature cum eo quod demonstratur, sed quod demonstratur in demonstratione est passio, ergo medium in demonstratione debet esse eiusdem nature cum passione; sed tale non potest esse nisi definitio passionis; ergo etc.

1.4 Item, medium in demonstratione debet dicere quid est et propter quid; sed tale est definitio passionis, dicit enim quid passionis prioris et propter quid passionis posterioris; ergo etc.

Et iste omnes fere sunt rationes Egidii.[b]

2.1 Oppositum arguitur: illud quod demonstratur debet demonstrari per causam alteram, quia Philosophus[c] dicit quod demonstratio est eorum quorum est causa altera, sed passio est illud quod demonstratur, ergo demonstratur per aliam causam; sed definitio passionis non est alia causa a passione essentialiter et realiter; ergo etc.

2.2 Item, in demonstratione potissima non debet peti principium; sed si in demonstratione medium esset definitio passionis, peteretur principium,

---

[a] Aristoteles, *AnPo*, II, 2, 90 a 5–11   [b] Egidius Romanus, *Super librum Posteriorum*, ad II, 9 (1488, f. 117D)   [c] Aristoteles, *AnPo*, I, 2, 71 b 20–25; I, 24, 85 b 24–27

---

4 definitio] subiecti *add. sed exp.* M   8 secundum rationem] *spatium vacans fere tres litterarum* M   19 fere] *iter.* M

Question 47.

Next we inquire concerning the middle term of a demonstration, and in the first place, whether the definition of the attribute is the middle term in a demonstration, in the second place, whether the definition of the subject is.

1.1 It appears that the definition of the attribute is, for the middle term in a demonstration, as the Philosopher says, is the account of the first term, but the first term in a demonstration is the major extreme and the attribute, therefore etc.

1.2 Furthermore, the middle term in a demonstration must be the middle term according to nature and not only according to reason; therefore that will be the middle term which is prior to the minor extreme, and posterior to the major extreme. But such is the definition of the attribute, for it is posterior to the major extreme, namely the attribute, and prior to the minor extreme, the subject. Therefore etc.

1.3 Furthermore, the middle term in a demonstration must be of the same nature as that which is demonstrated, but what is demonstrated in a demonstration is the attribute. Therefore the middle term in a demonstration must be of the same nature as the attribute. But there can be no such thing except for the definition of the attribute. Therefore etc.

1.4 Furthermore, the middle term in a demonstration must indicate what-it-is and because-of-what something is, but such is the definition of the attribute, for it indicates what the prior attribute is, and because-of-what the posterior attribute is. Therefore etc.

All of these are close to the arguments of Giles of Rome.

2.1 On the other hand, it is argued that that which is demonstrated must be demonstrated through some other cause, for the Philosopher says that demonstration is of that of which there is some other cause, but the attribute is what is demonstrated, therefore it is demonstrated through another cause. But the definition of the attribute is not a cause that is other than the attribute essentially or in actual reality, therefore etc.

2.2 Furthermore, in a demonstration of the strongest sort the question must not be begged, but if in a demonstration the middle term were the

quia in minori definitio passionis predicatur de subiecto, in conclusione predicatur passio de subiecto; ergo demonstratur passio de subiecto per hoc quod definitio passionis inest subiecto. Eque autem notum est quod passio insit subiecto et quod definitio passionis, quare peteretur principium.

3 Ad evidentiam istius questionis intelligendum quod tria sunt | in demonstratione, scilicet illud quod demonstratur et illud per quod demonstratur et illud de quo demonstratur aliud. Illud de quo demonstratur est subiectum, per quod est medium, et illud quod demonstratur est accidens vel passio. Sed accidentium sunt duo genera secundum duplicem naturam subiecti, scilicet materie et forme; unde quedam sunt accidentia que consequuntur totum aggregatum per formam, quedam autem per materiam. Et quia forma est principium speciei, materia autem individuationis, ideo illa que sequuntur totum per formam dicuntur accidentia consequentia speciem, que autem sequuntur ipsum totum per materiam dicuntur accidentia pertinentia ad individuum.

M 109rb

Iterum, quia forma est principium speciei et individua accidunt speciei ut sic, ideo accidentia consequentia formam vel speciem dicuntur accidentia per se, que autem sequuntur individuum dicuntur accidentia per accidens, quia non necessario insunt subiectis neque per se, et ideo sunt duplicia accidentia, scilicet accidentia per se et accidentia per accidens. Et in hoc conveniunt omnes quod accidentia [per accidens] non sunt demonstrabilia de subiectis nisi intelligatur de accidentibus ⟨per se. De accidentibus⟩ per accidens falsum est †quia cuiuslibet accidentis per accidens est per se subiectum†; omne enim demonstrabile demonstratur per causam necessariam in subiectis; talia autem non habent causam necessariam in subiectis; ideo etc. Accidentia autem necessaria et per se sunt demonstrabilia, et in ipsis invenitur ordo; inveniuntur enim quedam prima, quedam media, quedam ultima. Et cum talis diversitas sit in accidentibus, quidditas subiecti non est nisi causa unius accidentis primo et per se. Quidditas tamen diversimode considerata bene potest esse causa plurium accidentium, ut quidditas hominis in eo quod animal causa est sentiendi, secundum autem quod rationalis causa risibilis. Et ideo cum natura subiecti sit quid unum et determinatum, ab ipsa non est

---

30 et] *corr. ex* per M ‖ quare] *s. l.* M 33–34 demonstratur] scilicet illud *add. sed exp.* M 46 quia] que M 49 accidentibus] subiectibus M 50 est] quia cuiuslibet accidentis per accidens est per se *add. sed corruptum censemus* M 60 est] sunt (determinate *add. sed del.*) M

definition of the attribute the question would be begged, for the definition of the attribute is predicated of the subject in the minor premise, and in the conclusion the attribute is predicated of the subject. Therefore the attribute is demonstrated of the subject through this, that the definition of the attribute inheres in the subject. But it is known equally that the attribute inheres in the subject and that the definition of the attribute does so. Hence the question is begged.

3 To make the reply to this question evident, it should be understood what the terms are in a demonstration, namely that which is demonstrated, that through which it is demonstrated, and that of which the other is demonstrated. That of which the other is demonstrated is the subject, that through which it is demonstrated is the middle term, and that which is demonstrated is the accident or attribute. But there are two kinds of accidents, in accord with the twofold nature of the subject, namely, its matter and form. There are, then, some accidents that follow from the whole aggregate through its form, and others that follow from its matter. And since form is the principle of the species, and matter the principle of individuation, therefore those that follow from the whole through its form are called accidents that follow from the species, but those that follow from the whole through its matter are called accidents pertaining to the individual.

Furthermore, since form is the principle of the species, and individuals are accidental to the species as such, therefore accidents following from the form or species are called accidents *per se*. In contrast, those that follow the individual are called accidents *per accidens*, because they do not inhere in their subjects necessarily, nor *per se*. And therefore accidents are two-fold, namely accidents *per se* and accidents *per accidens*. And everyone agrees on this, that accidents are not demonstrable of their subjects unless one means accidents *per se*. For accidents *per accidens* this is false, since every demonstrable is demonstrated through a necessary cause in the subject, but accidents such as these do not have a necessary cause in their subjects, therefore etc. On the other hand, necessary and *per se* accidents are demonstrable, and in these an order is found. For some are found to be first, some intermediate, and some last. And since there is this sort of diversity in accidents, the quiddity of the subject is nothing but the cause of one accident primarily and *per se*. But the quiddity considered in different ways can very well be the cause of several accidents, as for instance the quiddity of human being, inasmuch as it is animal, is the cause of sensing, and insofar as it is rational it is the cause of the ability

immediate per se et primo nisi unum, et per illud aliud, et sic usque ad aliquem terminum. Et ideo bene ponitur ordo, quod quedam sunt prima, quedam media, quedam ultima.

Et hoc dicit Egidius[a] de hac materia, quia inter passiones causalitatem habentes quedam sunt prime, que sunt medium demonstrandi sed non demonstrantur, cum sint immediata et immediatorum non est demonstratio; alie autem sunt medie, que concluduntur per passiones priores et sunt media ad demonstrandum alias, unde et demonstrantur et demonstrant; alie autem sunt posteriores, que solum demonstrantur per priores et non demonstrant. Et ideo, cum inter passiones ipsas sit ordo, est una passio que est prima et immediate procedens a subiecto, et hec est medium in demonstratione et hec est indemonstrabilis.

Et hoc probat sic, quia si prima passio sit demonstrabilis, aut demonstratur per se ipsam, aut per passionem priorem, aut per posteriorem, aut per subiectum, aut ⟨per⟩ quidditatem eius aut per accidentia per accidens. Non per passionem priorem, cum positum sit ipsam esse primam. Iterum nec per se ipsam, quia tunc esset petitio in demonstratione. Nec per passionem posteriorem, quia illud per quod demonstratur aliquid in demonstratione potissima debet esse causa illius quod demonstratur; nulla autem passio posterior potest esse causa passionis prioris; quare per ipsam non demonstratur. Item, nec per accidentia per accidens, quia demonstratio non respicit talia accidentia. Item, nec per subiectum ipsum, tunc enim peteretur principium in tali demonstratione ut in maiori; si enim accipiatur medium ⟨ut⟩ subiectum in maiore, predicatur passio de subiecto, et item in conclusione, quare peteretur principium. Item, nec per quidditatem vel esse essentie subiecti, quia dicit Philosophus[b] quod demonstrationes non augentur per media, immo tanta debet esset proportio inter conclusionem et medium in demonstratione potissima quod tantum debet esse unum medium ad unam conclusionem et e converso; manifestum autem quod per quidditatem subiecti demonstratur alia passio posterior; quare per ipsam non demonstratur passio prima. Item, nec per esse actuale subiecti, quia tale esse respicit accidentia per accidens; talia autem non respicit demonstratio; quare etc. Quare ipsa prima passio est indemonstrabilis et ideo ipsa est medium in demonstratione potissima.

[a] Egidius Romanus, *Super librum Posteriorum*, ad II, 9 (1488, f. 117D) [b] Aristoteles, *AnPo*, I, 12, 78 a 15

---

63 media] ultima M   67 priores] possibilis M   69 posteriores] possibilis M   85 peteretur] *iter.* M

to laugh. And since the nature of a subject is some one and determinate thing, there thus comes from it no more than one immediate ⟨accident⟩, *per se* and primary, and through that another, and so on up to the one that is final. And therefore, it is correct to posit an order, because some are first, some intermediate, and some last.

And this is what Giles of Rome says of this matter, that between attributes having causality some are primary, those being the middle terms of demonstrating, but they are not demonstrated since they are immediate and there is no demonstration of what is immediate. Others are middle terms which are reached as a conclusion through prior attributes and are themselves middle terms for demonstrating others, so that they both are demonstrated and demonstrate. Still others are posterior attributes which are only demonstrated through prior ⟨attributes⟩ and do not themselves demonstrate. Therefore, since there is an order among these attributes, there is one attribute that is first and that proceeds immediately from the subject. And this is the middle term in a demonstration, and it is indemonstrable.

He proves this as follows. If the first attribute is demonstrable it is demonstrated either through itself, through a prior attribute, through a posterior attribute, through its subject, through its quiddity, or through its accidents *per accidens*. It will not be demonstrated through a prior attribute, since it is posited that it is itself the first. Nor will it be demonstrated through itself, since then there would be a begging of the question in the demonstration. Nor will it be demonstrated through a posterior attribute, since that through which something is demonstrated in a demonstration of the strongest sort must be the cause of what is demonstrated, but no posterior attribute can be the cause of a prior attribute, and therefore it is not demonstrated through a posterior accident. Further, it is not demonstrated through accidents *per accidens*, since there is no demonstration with respect to such accidents. Further, it is not demonstrated through its subject, for then the question is begged in such a demonstration, that is, in the major premise. For if the middle term is taken as the subject in the major premise, then the attribute is predicated of the subject in the major premise, and again in the conclusion, so that the question is begged. Again, it is not demonstrated through the quiddity or being of the essence of the subject, for the Philosopher says that demonstrations are not augmented through their middle terms. Instead, the proportion between the conclusion and the middle in a demonstration of the highest sort must be such that there is only one conclusion for one middle term, and conversely.

Contra istud aliquis argueret: quod dicitur primo, quod passio prima est indemonstrabilis de subiecto, diceret aliquis quod falsum est. Et ratio huius est ista: manifestum est enim quod omnis passio que per se inest subiecto, sive fuerit prima sive mediata sive postrema, est diversa a subiecto, ex hoc quod per se inest subiecto, causam habet in subiecto; sed omne quod est diversum a subiecto et causam habet in subiecto potest demonstrari de subiecto; quare omnis passio, sive prima sive media sive ultima, demonstrabilis est de subiecto. Nec potest dici quod prima passio non habet causam in subiecto suo, quia omnis passio quantumcumque prima causatur a principiis essentialibus subiecti; quare causam habet in subiecto illo per quam poterit demonstrari.

Item, in per se accidentibus rei omne illud est demonstrabile de quo contingit querere propter quid, ex quo enim contingit querere propter quid, causam habet, | et si causam habet, ergo est demonstrabile; sed de omni accidente quantumcumque primo contingit querere propter quid inest subiecto, ex hoc enim quod in intellectu hominis non includitur risibile, et ei inest per se, contingit querere propter quid sibi inest; quare erit demonstrabile.

M 109va

Et item in dictis videtur importari contradictio, si ponamus quod esse in effectu sit res addita ipsi rei, sicut ponit, dato quod sit passio et immediate tamen consequitur subiectum, materia enim cum forma causa est primo ipsius esse, et per consequens omnium aliorum, quia omnia accidentia presupponunt esse in effectu; et tamen ipse ponit quod esse in effectu potest demonstrari de subiecto suo ut de ente; quare prima passio potest demonstrari de suo subiecto.

Ad rationes igitur in oppositum.

'Si demonstratur aut demonstratur per priorem etc.', dico quod per nullum horum, sed demonstratur per subiectum et quidditatem subiecti. Et

---

113 importari] importare M   121 si ... etc.] *Cf.* 2.2   122 subiecti] *iter.* M

But it is obvious that through the quiddity of the subject another, posterior attribute is demonstrated, and so the first attribute is not demonstrated through that quiddity. Further, neither is it demonstrated through the actual being of the subject, for such being is with respect to accidents *per accidens*, and there is no demonstration with respect of these, therefore etc. Therefore, this first attribute is indemonstrable, and thus it is the middle term in the highest sort of demonstration.

Against this, some have argued that what is claimed first—that the first attribute is indemonstrable of its subject—would be said by some to be false. The reason is that it is obvious that every attribute that inheres in a subject *per se*—whether it is first, intermediate, or last—is distinct from that subject. Because it inheres *per se* in the subject, it has a cause in the subject. But whatever is distinct from a subject and has a cause in the subject can be demonstrated of that subject, and so every attribute, whether first, intermediate, or last, is demonstrable of its subject. Nor can it be said that the first attribute does not have a cause in its subject, since every attribute, even the first, is caused by the subject's essential principles, so that it has a cause in the subject through which it can be demonstrated.

Furthermore, in the *per se* accidents of a thing, everything is demonstrable of which one can ask because-of-what, since if it can be asked because-of-what then it has a cause, and if it has a cause then it is demonstrable. But of every accident, even the first, it can be asked because-of-what it inheres in the subject. So, since the ability to laugh is not included in the concept of a human being, and yet it inheres in a human being *per se*, one can ask because-of-what the ability to laugh inheres in a human being, and so it will be demonstrable.

Furthermore, there seems to be a contradiction in what has been claimed, if we posit that existing in actual effect is a thing added to the thing itself, as Giles assumes, on the supposition that it is an attribute and is immediate, and yet that it follow from its subject. For matter along with form is the primary cause of this existence, and consequently the cause of all the rest, since all accidents presuppose existence in actual effect. And yet he posits that existence in actual effect can be demonstrated of its subject as of a being. As a result, the first attribute can be demonstrated of its subject.

To the arguments for the opposite position.

To Giles's argument that 'If it is demonstrated, it is demonstrated either through something prior, etc.'. I reply that it is demonstrated through

tu dicis: unum medium tantum erit in unam conclusionem; et per quidditatem subiecti demonstratur esse actuale de subiecto; quare etc. Verum est, tantum unum medium erit ad unam conclusionem eque primo, non est tamen inconveniens quod unum medium sit ad multas conclusiones non eque primo. Et ideo dicerem quod per quidditatem subiecti demonstratur passio prima de subiecto; item, esse actuale de subiecto demonstratur, sed non primo.

Vel posset dici quod licet esse actuale sit passio consequens subiectum, non tamen talis passio est que possit ostendi de subiecto, quia illa passio proprie demonstratur de subiecto demonstratione potissima et per quidditatem subiecti que universaliter inest subiecto, universalitate determinata in hoc libro. Sed esse actuale nulli subiecto inest universaliter preter quam primo principio; inesse enim universaliter est quando solum inest illi subiecto, et si insit aliis, inerit illis per naturam illius subiecti. Nunc autem esse in effectu inest aliis ab homine, et item etsi aliis insit non inest illis per naturam hominis, et ideo esse actuale non potest demonstrari de homine. Verumtamen verum est quod esse actuale potest demonstrari de subiecto, sicut dicit Philosophus[a] post, sed hoc est inquantum diversum est secundum rationem a subiecto, et non inquantum est idem essentialiter cum subiecto.

Dicendum igitur quod definitio passionis non est medium in demonstratione potissima, et huius causa apparet ex precedentibus; nihil enim est medium in demonstratione potissima quod est demonstrabile. Sed universaliter omnis passio, sive prima sive media sive ultima, est demonstrabilis de subiecto, et per consequens ipsarum definitio, quia eadem ratio est de passione et de eius definitione, cum idem sint essentialiter; quare neque passio neque definitio passionis erit medium.

Ad cuius evidentiam considerandum est quod definitio passionis duplex est: quedam dicens quid est nomen, ponens altera nomina, quedam autem dicens quid est ipsa passio, et illa duplex, quia una est que [subiectum] accipit et principia formalia et essentialia passionis in genere suo, et alia que accipit subiectum in definitione sua, et illa est completissima inter

[a] Aristoteles, *AnPo*, II, 7, 92 b 12–18

---

124 demonstratur] determinatur M   130 licet] habet M || sit] sicut M   139 verum] *perperam del.* M   142 idem] universaliter *add. sed del.* M   146 passio] demonstratio M   147 ipsarum] ipsorum M

none of these, but rather through the subject *and* the subject's quiddity. You reply that 'there will be one middle term only for one conclusion, and the actual existence of the subject is demonstrated through the subject's quiddity; therefore etc.'. It is true that there will be only one middle term for one equally primary conclusion, but it is not absurd for there to be one middle term for several conclusions that are not equally primary. And therefore I would say that the first attribute is demonstrated of the subject through the subject's quiddity. Further, actual existence is demonstrated of the subject, but not primarily.

Alternatively, it can be said that although actual existence is an attribute that follows the subject, there is no such attribute that can be established of the subject, for that attribute is strictly demonstrated of the subject by a demonstration of the strongest sort, and through the quiddity of the subject, which inheres in the subject universally through the universality set out in this book. But actual existence is in no subject universally except for the first principle. For inhering universally occurs when it inheres only in that subject, and if it inheres in others it inheres in them through the nature of that subject. But existing in actual effect inheres in others besides human beings, and again, although it inheres in others it does not inhere in them through the nature of human being. Therefore actual existence cannot be demonstrated of human being. Nonetheless it is true that actual existence can be demonstrated of a subject, as the Philosopher says later, but this is so insofar as it is different in its account from that subject, not insofar as it is essentially the same as its subject.

It should be said, therefore, that the definition of an attribute is not the middle term in a demonstration of the strongest sort, and the cause of this is apparent from what has gone before. For nothing that is demonstrable is the middle term in a demonstration of the highest sort. But universally, every attribute, whether first, intermediate, or last, is demonstrable of its subject, and consequently the definition of these attributes is demonstrable, since the same account holds for the attribute and its definition, since they are the same essentially. Therefore neither the attribute nor the definition of the attribute will be a middle term.

To make this evident, one should consider that the definition of an attribute is two-fold: on the one hand it indicates what the name is, by introducing other names, and on the other hand it indicates what the attribute itself is. This latter is itself two-fold; one kind takes the formal and essential principles of the attribute in its genus; the other takes the

omnes. Manifestum autem quod definitio dicens quid significat nomen non potest esse medium, quia ipsa non est causa sufficiens passionis ad subiectum; immo si esset causa sufficiens, cum non entium sit definitio ponens altera nomina, ipsorum posset esse demonstratio, quod negat Philosophus. Item, nec definitio accipiens principia essentialia passionis, quia ista definitio erit essentialiter eadem cum passione; ostensum autem est quod omnis passio demonstratur de subiecto vel est demonstrabilis de ipso. Item, nec definitio accipiens principia sui subiecti, quia illud quod est tota demonstratio positione differens non potest esse medium; sed talis definitio passionis est tota demonstratio, accipit enim passionem et subiectum et quod quid est subiecti et quod quid est passionis; quare non potest esse medium in demonstratione.

Licet tamen in demonstratione potissima definitio passionis non sit medium, tamen in aliqua demonstratione bene est medium, quia, ut suppono, medium in demonstratione est definitio subiecti vel primi termini; nunc autem possibile est quod una passio demonstretur de alia; et ideo definitio talis passionis prioris potest esse medium talis demonstrationis, illa enim definitio dicit quid passionis prime sed propter quid posterioris. Unde bene est invenire passionem demonstrabilem et demonstrantem, similiter passionem demonstratam solum, sed non possumus reperire passionem demonstrantem solum.

Ad 1.1 Ad primum ergo, cum dicitur quod primus terminus est passio, dico quod falsum est, sed ipsum subiectum vel quod quid est ipsius subiecti, quia dicit Philosophus[a] in I *Physicorum* quod subiectum est prius predicato. Tamen licet hoc non diceret, bene diceretur primus terminus; illud enim dicitur esse primus terminus quod secundum causalitatem precedit omnia alia in demonstratione et omne illud est medium dicens quod quid est subiecti, dicit enim quid est subiectum | et propter quid [est] passionis, et ideo ipsum quod quid est subiecti vel subiectum est ratio primi termini.

M 109vb

---

[a] Aristoteles, *Phys.*, I, 6 189 a 31–32

156 non²] *i. m.* M   172 prime] prima *ante corr.* M   182 propter] quod M

subject in its definition, and this last is the most complete of them all. Now it is obvious that the definition indicating what the name signifies cannot be the middle term, since it is not a sufficient cause of the attribute in relation to the subject. Indeed, if it were a sufficient cause, then since for non-beings there is a definition that introduces other names, there could be a demonstration of non-beings, which the Philosopher denies. Further, the definition that takes the essential principles of the attribute is also not the middle term, since this definition will be the same in essence with the attribute. But it has been shown that every attribute is demonstrated of its subject or is demonstrable of it. Further, the definition that takes the principles of its subject is also not the middle term, since that which is the whole demonstration, differing only in how it is set out, cannot be the middle term. But such a definition of the attribute is the whole demonstration, for it takes the attribute, the subject, the essence of the subject, and the essence of the attribute. Hence it cannot be the middle term in a demonstration.

Now although the definition of the attribute is not the middle term in the strongest sort of demonstration, still it is certainly the middle term in some demonstrations, since, as I suppose, the middle term in a demonstration is the definition of the subject or the first term. But it is possible for one attribute to be demonstrated of another, and therefore the definition of such a prior attribute can be the middle term in such a demonstration. For that definition indicates the what-it-is of the first attribute, and the because-of-what of the second attribute. Hence one may very well find a demonstrable and demonstrating attribute, and similarly, an attribute that is demonstrated alone, but one cannot find an attribute that demonstrates alone.

Ad 1.1 In response to the first argument, when it is claimed that the first term is the attribute, I reply that this is false, but the subject, or the subject's essence, is the first term, for the Philosopher says in *Physics* I that the subject is that which is predicated beforehand. But even if he had not said this, it would have been suitable to call ⟨the subject or its essence⟩ the first term. For that is said to be the first term which, according to causality, precedes all others in demonstration, and every such thing is a middle term indicating the essence of the subject. For it says what the subject is, and because-of-what the attribute is, and therefore the essence of the subject, or the subject, is the account of the first term.

Ad 1.2 Ad aliud, 'medium in demonstratione debet esse medium natura', verum est quod illud debet esse recte medium per quod accipitur cognitio et scientia simpliciter. Certe in his que sunt secundum naturam debet accipi medium quod est medium secundum naturam; in illis autem que sunt secundum rationem et intellectum oportet accipere medium in cognoscendo et in intelligendo. Et ideo, cum demonstratio sit actus rationis, manifestum quod debet accipi tale medium quod est prius in cognoscendo et medium recte in cognoscendo. Illud autem est illud de quo notior est ipsa passio quam ipsa passio de subiecto quod significatur per conclusionem [vel de quo subiecto] et de quo ipsum subiectum est notius quam ipsa passio de subiecto. In demonstratione enim sic procedimus, per hoc enim quod medium inest subiecto concludimus passionem inesse subiecto, quod medium erit tale quod est notius de subiecto quam ipsa passio. Et item in demonstratione sic procedimus quod passio demonstratur de subiecto per hoc quod passio prius inest ipsi medio. Et manifestum est quod passio notior est de medio quam sit de subiecto. Rectissime ergo medium tale ut quod quid est subiecti est medium in cognoscendo et in intelligendo.

Ad 1.3 Ad tertium, cum arguitur: medium debet esse eiusdem nature cum eo quod demonstratur, dico quod falsum est, quia Philosophus dicit quod illud quod demonstratur demonstratur per causam aliam; causa autem alia diversa est per essentiam a causato; et ideo medium debet esse diversum ab eo quod demonstratur. Et non requiritur identitas secundum rem, sed proportio que est inter causam et causatum. Et cum dicitur 'passio est illud quod demonstratur', verum est, et ideo medium ipsius erit aliud a passione; erit enim definitio ipsius subiecti quod est causa talis passionis et inherentie ipsius ad subiectum.

Ad 1.4 Ad quartum, dicendum quod illud quod dicit quid et propter quid est definitio subiecti; dicit enim quid subiecti et propter quid inherentie passionis ad subiectum, et ideo per istam rationem non potius concluditur quod medium sit quod quid est passionis quam ipsius subiecti.

---

188 medium[1]] per quod accipitur cognitio et scientia simpliciter certe in his que sunt secundum naturam debet accipi medium *add.* M 203 tertium] quartum M 210 erit enim] et non M 203–215 Ad tertium cum arguitur … quam ipsius subiecti] *responsiones inv.* M 212 quartum] tertium M 213 inherentie] inherentia M

Ad 1.2 In response to the second argument, 'the middle term in a demonstration ought to be the middle term by nature', it is true that that ought rightly to be the middle term through which cognition and knowledge said without qualification is received. Certainly in those demonstrations that proceed in accord with nature one ought to take as the middle term that which is the middle term according to nature. But in those demonstrations that proceed according to reason and intellect one ought to take the middle term in cognizing and understanding. And therefore, since demonstration is an act of reason, it is clear that one ought to take a middle term that is prior in cognizing and that is correct in cognizing. That, however, would be something whose attribute is better known than is the attribute of the subject that is signified in the conclusion, and whose subject is better known than is the attribute of the subject. For in demonstration we proceed as follows: through the fact that the middle term inheres in the subject, we conclude that the attribute inheres in the subject, which middle term will be such that it is better known of the subject than is the attribute. And, moreover, we proceed in demonstration in such a way that the attribute is demonstrated of the subject through the fact that the attribute is prior in its inherence in the middle term. And it is clear that the attribute is better known of the middle term than it is of the subject. It is most correct, then, that a middle term such as the essence of the subject is the middle term in cognizing and understanding.

Ad 1.3 In response to the third, when it is argued that the middle term must be of the same nature as what is demonstrated, I reply that this is false. For the Philosopher says that what is demonstrated is demonstrated through some other cause, but another cause is different in essence from what it causes, and therefore the middle term must be different from that which is demonstrated. What is required is not real identity, but the proportion that holds between cause and what is caused. And when it is said that what is demonstrated is the attribute, this is true, and therefore the middle term of this demonstration will be something other than the attribute. For it will be the definition of the subject, which is the cause of such an attribute and of its inherence in its subject.

Ad 1.4 In response to the fourth, it should be held that that which indicates the what-it-is and the because-of-what is the definition of the subject. For the definition indicates the what-it-is of the subject, and it indicates because-of-what there is an inherence of the attribute to subject. Therefore this argument no more concludes that the middle term is the essence of the attribute than it concludes that it is the essence of the subject.

⟨Questio 48⟩

Consequenter queritur utrum definitio subiecti sit medium in demonstratione.

1.1 Et arguitur quod non, quia medium in demonstratione debet esse unigeneum cum passione probanda; sed quod quid est subiecti non est unigeneum cum passione probanda; ergo etc.

1.2 Item, in demonstratione medium debet esse notius subiecto, quia per medium debet demonstrari passio de ipso; sed quod quid est subiecti non est notius subiecto, quia est idem cum subiecto; ergo etc.

2 Oppositum arguitur, quia definitio subiecti medium est in syllogismo qui procedit ex primis, veris et immediatis; sed talis syllogismus est demonstratio; ergo definitio subiecti medium est in demonstratione.

3 ⟨Quidam dicebant quod definitio subiecti non potest esse medium in demonstratione⟩, tum quia si definitio subiecti esset medium in demonstratione, demonstratio non procederet ex propriis, tum quia demonstratio non procederet ex immediatis, tum quia hoc dato, premisse non confirmarentur per primum principium.

Primum declarabant sic, quia quidditas subiecti continet in se virtualiter omnes passiones et proprietates que fluunt ab essentia subiecti. Sicut enim in ipsa arte in quodam principio communi continentur virtualiter omnia illa que possunt produci per artem illam, ita quod [sicut] ars demonstrandi communis est omnibus illis que possunt produci per artem illam, sic in quidditate subiecti continentur omnes passiones que possunt produci ab illa, ita quod ipsum subiectum sit commune omnibus passionibus que fluunt ab essentia subiecti. Quare si est aliqua demonstratio procedens ex medio quod est quod quid est subiecti, procedet ex communibus.

Item, non procederet ex immediatis, quia manifestum est quod ⟨ab⟩ ipsa quidditate subiecti non fluunt omnes passiones et proprietates subiecti,

---

4 passione] demonstratione M ‖ non] *iter.* M   5 passione] demonstratione M   8 subiecto²] medio M

Question 48.

Next we ask whether the definition of the subject is the middle term in a demonstration.

1.1 It is argued that it is not, since the middle term in a demonstration must be of the same genus as the attribute to be proved. But the essence of the subject is not of the same genus as the attribute to be proved. Therefore etc.

1.2 Furthermore, in a demonstration the middle term must be better known than the subject, since the attribute must be demonstrated of the subject through the middle term. But the essence of the subject is not better known than the subject, since it is the same as the subject. Therefore etc.

2 The opposite is argued for, because the definition of the subject is the middle term in a syllogism that proceeds from things that are first, true and immediate, and such a syllogism is a demonstration, therefore the definition of the subject is the middle term in a demonstration.

3 Some have said that the definition of the subject cannot be the middle term of a demonstration. One reason is that, if the definition of the subject were the middle term in a demonstration, then the demonstration would not proceed from things that are proper. A second reason is that then the demonstration would not proceed from immediate propositions. A third is that, if this were the case, then the premises would not be confirmed through a first principle.

The first they have set out as follows. The quiddity of the subject contains in itself virtually all attributes and properties that flow from the essence of the subject. For just as in an art a certain common principle contains virtually everything that can be produced through the art, with the result that the art of demonstrating is common to everything that can be produced through the art, so too in the quiddity of the subject are contained all the attributes that can be produced from it, with the result that this subject is common to all the attributes that flow from the essence of the subject. Therefore if some demonstration proceeds from a middle term which is the essence of the subject, it proceeds from things that are common rather than proper to it.

Second, they argue that the demonstration would not proceed from immediate propositions, since it is obvious that not all the attributes and

sed solum una, et mediante illa omnes alie, usque ad aliquem terminum. Si igitur contingeret per quidditatem subiecti demonstrare omnes passiones, cum non omnes fluant ab ipsa immediate, sequeretur quod demonstratio non procederet ex immediatis.

Item, sequeretur quod premisse demonstrationis non confirmarentur per primum principium, quia in demonstratione tali in minore predicatur quod quid est subiecti de subiecto; quod quid est autem subiecti est idem cum subiecto, quare, cum ipsa sit immediata, non erit alia prior; quare non confirmatur per primum principium.

Ista non videntur habere veritatem. Quod enim ipsi dicunt primo, quod si definitio subiecti esset medium in demonstratione, quod demonstratio procederet ex communibus, ⟨falsum est⟩, quoniam in tali demonstratione proceditur ex propriis, quia illa demonstratio potissime procedit ex propriis in qua medium est appropriatum maxime ei quod demonstratur et ei de quo demonstratur. Sed in demonstratione in qua sumitur pro medio quidditas subiecti, medium appropriatur passioni, cum passio prima immediate fluat ab essentia et quidditate subiecti, hoc autem non esset nisi quidditas subiecti esset appropriata sibi. Et iterum, ad | utrumque comparatur in aliquo genere cause, quia comparatur ad passionem sicut causa efficiens eius et ad subiectum sicut causa formalis eius, quia quod quid est subiecti est causa accidentium et dicit totam naturam subiecti, non plus nec minus.

Item, manifestum est quod ipsi falsum dicunt secundo, quod si quod quid est subiecti sumeretur pro medio, demonstratio non procederet ex immediatis, quia illa demonstratio procedit ex immediatis que accipit premissas immediatas; sed in demonstratione ubi quod quid est subiecti est medium, in maiore comparatur passio ad quod quid est subiecti ut effectus immediatus ad causam, et manifestum est quod ei immediate inest et ita est quartus modus dicendi per se, in minori autem quod quid est subiecti comparatur ad subiectum ut quod quid est ad illud cuius est; manifestum est ergo quod premisse sunt immediate.

---

45 appropriata] ap(ro)por^ta M   46–47 sicut ... eius] cause efficientis M   55 immediatus] immediatur M

properties of the subject flow from the quiddity of the subject, but only one, and by means of it all the others flow, up to the last. If, therefore, it were possible through the quiddity of the subject to demonstrate all the attributes, then since they do not all flow immediately from this quiddity, it would follow that demonstration would not proceed from immediate propositions.

Further, they argue that it would follow that the premises of the demonstration would not be confirmed through a first principle, since in such a demonstration the essence of the subject is predicated of the subject in the minor premise. But the essence of the subject is the same as the subject, so, since it is immediate, there will not be anything prior, so that it is not confirmed through a first principle.

These claims do not seem to have any truth. As for what they say first, that if the definition of the subject is the middle term in a demonstration then the demonstration would proceed from common terms, this is false, since in such a demonstration one proceeds from proper terms. For a demonstration of the strongest sort proceeds from proper terms in which the middle term is most proper to that which is demonstrated, and to that by which it is demonstrated. But in a demonstration in which one takes the quiddity of the subject as the middle term, the middle is proper to the attribute, since the first attribute flows immediately from the essence and quiddity of the subject, but this would not be the case unless the quiddity of the subject were proper to it. Moreover, it is related to each in some genus of cause, since it is related to the attribute as its efficient cause and to the subject as its formal cause, since the essence of the subject is the cause of its accidents, and indicates the whole nature of the subject, neither more nor less.

Moreover, it is obvious that they say something false on the second count, that if the essence of the subject were taken as the middle term then the demonstration would not proceed from immediate propositions. For a demonstration proceeds from immediate propositions when it takes on immediate premises. But in a demonstration where the essence of the subject is the middle term, in the major premise the attribute is related to the essence of the subject as an immediate effect to its cause. And clearly, it inheres in it immediately, and so this is the fourth mode of speaking *per se*. But in the minor premise the essence of the subject is related to the subject as essence to that of which it is the essence. It is obvious, then, that the premises are immediate.

Et tunc ad formam rationum, quando ipsi arguunt primo 'in ipso subiecto virtualiter continentur passiones etc.', dico quod secundum quod Philosophus[a] vult II *De generatione,* idem manens idem est causa eiusdem per se et primo. Verum est quod quidditas subiecti secundum suam rationem considerata est una, et ideo per se et primo est causa unius passionis. Unde licet quidditas sit communis pluribus operationibus et proprietatibus, tamen appropriatur uni proprietati, sicut quidditas hominis est communis ut ab ipsa procedant iste operationes vegetare, sentire, ratiocinari, tamen ratiocinari per se et primo procedit ab ipsa, et ideo quidditas hominis appropriata est ei quod est ratiocinari.

Similiter quod dicunt ulterius quod demonstratio non procederet ex immediatis, dico quod per quidditatem subiecti non demonstratur nisi una passio de subiecto per se et primo, et talis demonstratio est immediata, quia illa passio immediate causatur a principiis subiecti. Unde si per quidditatem subiecti demonstratur passio talis, illa est demonstratio ex immediatis, si autem demonstretur alia passio mediata, manifestum est quod talis demonstratio procedit ex premissis mediatis.

Et quod ipsi dicunt tertio, quod premisse demonstrationis non confirmarentur per primum principium, hoc est falsum, quia rationes cuiuslibet termini manifestantur nobis ex terminis communibus, ut entis; ergo rationabile ⟨est⟩ quod omnes propositiones que formantur ex terminis specialibus manifestarentur ex propositione que sumitur sive formatur ex terminis communibus, ut entis, ut ex ista 'ens est ens'. Et ideo quantumcumque predicatur idem de se vel quocumque aliquo alio modo in terminis specialibus, nihilominus omnes habent unitatem ex primo intelligibili quod formatur in terminis entis, quoniam secundum Commentatorem[b] ens est illud quod primo occurrit intellectui.

Dicendum est igitur alio modo, videlicet quod definitio subiecti medium est in demonstratione. Et ad huius evidentiam considerandum est quod in omni subiecto de quo demonstratur aliqua passio, nos consideramus duplicem naturam: materiam et formam. Et ideo, si sint alique forme per se existentes per naturam, de eis ut sic non demonstrantur alique passiones, propter quod scientie particulares, quarum proprie est demonstrare, non se extendunt ad tales substantias; unde metaphysicus unde talis non

[a] Cf. Aristoteles, *GC,* II, 10, 336 a 27–28   [b] *Rectius* Avicenna latinus, *De philosophia prima vel scientia divina,* I, 5 (ed. Van Riet, p. 35)

66 iste] omnes *ante corr.* M   67 ipsa] ipso M   72 si] *s. l.* M   82 modo] *s. l.* M   86 igitur] quod *add. sed exp.* M   88 nos] non *ante corr.* M

As for the form of their arguments, when they argue in the first place that the attributes are virtually contained in the subject etc., I reply that, according to what the Philosopher holds in *De generatione* II, the same remaining the same is the cause of the same *per se* and primarily. It is true that the quiddity of the subject considered according to its account is one, and therefore *per se* and primarily it is the cause of one attribute. Accordingly, even though the quiddity is common to many operations and properties, still it is proper to one property. For instance, the quiddity of human being is common in such a way that the operations of nutrition, sensation, and reasoning proceed from it. Still, reasoning proceeds from it primarily and *per se*, and therefore the quiddity of human being is proper to that which is reasoning.

As for their saying that demonstration would not proceed from immediate propositions, I similarly reply that nothing is demonstrated through the quiddity of the subject except one attribute of the subject, primarily and *per se*, and such a demonstration is immediate since that attribute is caused immediately from the principles of the subject. Hence, if such an attribute is demonstrated from the quiddity of the subject, such a demonstration is from immediate propositions, but if some other mediated attribute is demonstrated, then it is obvious that such a demonstration proceeds from mediate premises.

As for their third claim, that the premises of a demonstration are not confirmed through a first principle, this is false, since the accounts of any such term are made clear to us from common terms, such as 'being'. Therefore it is reasonable that all propositions formed from specific terms are made clear from a proposition which is drawn or formed from common terms, such as 'being', as for instance from 'being is being'. And therefore however much the same is predicated of itself or of anything else in any other way, in specific terms, they nevertheless all have their unity from the first intelligible, which is formed in terms of being, since according to the Commentator being is that which first occurs to the intellect.

And so a response should be made in another way, namely that the definition of the subject is the middle term in a demonstration. To make this evident it should be considered that in every subject of which attributes are demonstrated we consider a two-fold nature, matter and form. And therefore, if there are forms existing on their own through nature, then no attributes are demonstrated of these as such. For this reason the particular sciences, to which demonstration strictly belongs, do not extend to such

demonstrat, sed inquirit. Et ideo Commentator[a] II *Metaphysice* dicit quod mathematica sunt in primo gradu certitudinis, et ea sequuntur naturalia, metaphysicalia autem sunt minus certa quoad viam demonstrationis, cum eorum non sit proprie [demonstrare] demonstratio sed inquisitio simplex, cum non habeant passionem. Et propter hoc demonstrationes non inveniuntur nisi in illis que habent materiam et formam, et propter hoc habent duo genera accidentium secundum istas duas naturas; sunt enim quedam consequentia aggregatum per formam, quedam per materiam. Et quod quid est subiecti est causa omnium proprietatum subiecti, quia omnes proprietates sunt posteriores ipso quod quid; nunc autem ita est quod illud quod est prius est causa posteriorum, et ideo quod quid est subiecti est causa omnium proprietatum eius et passionum; quare quod quid est subiecti vere ponendum est medium. Et istius ultimi declaratio est, quia illud est medium in demonstratione potissima quod non habe[n]t causam aliam in subiecto et quod est causa omnium aliorum in subiecto; sed illud est quod quid est subiecti vel definitio subiecti, quia quod quid est subiecti non habet causam aliam in subiecto; fatuum enim est querere quare aliquid est suum quod quid est, ipsum enim est suum quod quid est, quia ipsum est ipsum, sicut fatuum est querere quare homo est homo.

Item, illud est medium in demonstratione potissima quo habito non contingit ulterius querere propter quid; sed habito quod quid est subiecti pro medio non contingit ulterius querere propter quid est, non ⟨enim⟩ queritur quare ipsum inest subiecto.

Item, subiecto inest quod quid est subiecti per se et non contingit querere quare subiecto inest, cum in subiecto sit causa predicati. Et hoc est de intentione Philosophi[b] I *De anima*; dicit enim ibi quod quod quid est subiecti est principium omnis demonstrationis, principium autem demonstrationis est medium demonstrationis.

Item, Philosophus[c] dicit secundo huius quod medium in demonstratione potissima dicit quid et propter quid; quod quid autem subiecti est illud quod dicit [quod] quid est subiecti et propter quid inherentie passionis | ad ipsum; manifestum est igitur, ut videtur, quod quod quid est ⟨subiecti⟩

M 110rb

---

[a] Averroes, *Commentarium magnum in Metaphysicam Aristotelis*, II, 3, comm. 16 (1562, f. 35K)
[b] Aristoteles, *De an.*, I, 1, 402 b 25–26   [c] Aristoteles, *AnPo*, II, 2, 90 a 32

---

94 ea] ens M   104–105 et ... medium] *post* homo M   116 subiecto] passio M   118 subiecti] subiectum *ante corr.* M

substances. Hence the metaphysician as such does not demonstrate, but only inquires. And therefore the Commentator says, on *Metaphysics* II, that mathematics holds the first degree of certainty, and natural entities follow. But the objects of metaphysics are less certain, as far as their path of demonstration is concerned, since of these there is not, strictly speaking, demonstration, but simple inquiry, since they have no attributes. Because of this, demonstrations are found only in those things that have matter and form, and because of this they have two genera of accidents in accord with these two natures. For there are certain inferences assembled through the form, and others through the matter. And the essence of the subject is the cause of all the properties of the subject since every property is posterior to its essence. Now that which is prior is the cause of what is posterior, and therefore the essence of the subject is the cause of all its properties and attributes, so that the essence of the subject is truly to be posited as the middle term. And the explanation of this last claim is that that is the middle term in a demonstration of the strongest sort which does not have any cause in the subject, and which is the cause of every other thing in the subject. But this is the essence of the subject, or the definition of the subject, since the essence of the subject does not have another cause in the subject. For it is foolish to ask why something is its essence, for it is its essence because it is itself. In this way, it is foolish to ask why a human being is a human being.

Further, that is the middle term in demonstration of the strongest sort which, when it is had, it is not possible to ask because-of-what any further. But when one has for one's middle term the essence of the subject it is not possible to ask because-of-what any further. For one does not ask why this essence inheres in the subject.

Furthermore, the essence of the subject inheres *per se* in its subject, and one does not ask why it inheres in the subject, since it is the cause of the predicate in the subject. This is the meaning of the Philosopher in *De anima* I. He says there that the essence of the subject is the principle of every demonstration, but the principle of a demonstration is the demonstration's middle term.

Further, the Philosopher says in the second book of this work that the middle term in a demonstration of the strongest sort indicates what-it-is and because-of-what. But that which a subject is is that which indicates the essence of the subject and that because-of-which the attribute inheres

125 est medium in demonstratione potissima.

Item, principium demonstrationis est medium. Quero de illo principio: aut est complexum aut incomplexum; non potest esse complexum, quia illud quod demonstratur et de quo demonstratur est incomplexum; erit ergo medium incomplexum. Si sic, aut erit quod quid est passionis aut quod quid est subiecti; non poterit esse quod quid est passionis, cum ipsum sit demonstrabile de subiecto; relinquitur ergo quod ipsum sit quod quid est subiecti. Et hoc est quod Philosophus[a] dicit IV *Physicorum*, capitulo de loco; dicit enim quod talis debet queri definitio loci ut ex ipsa reddatur quod quid est loci, et solvatur controversia opinionum, et quod per eam reddatur causa omnium passionum et proprietatum eius. Et ibi dicit Commentator[b] quod perfecte definitiones rerum nate sunt reddere causas omnium accidentium existentium in rebus illis; tale ergo debet esse medium in demonstratione ex quo reddatur causa omnium passionum. Hoc autem est definitio subiecti, cum ipsa per sua principia formalia sit causa omnium proprietatum et passionum sibi inherentia, et hoc est intelligendum in demonstratione potissima. Tamen verum est quod in demonstratione non potissima definitio passionis poterit esse medium demonstrandi unam passionem de alia, et ipsa iterum poterit esse medium ad demonstrandum [ad] aliam et ipsa solum demonstratur, non autem demonstrat.

Ad rationes.

Ad 1.1 Ad primam, dico quod non oportet quod medium sit unigeneum cum passione ita quod eiusdem nature, sed unigeneum illa unitate que est inter causam et causatum; talis autem unitas est inter quod quid est subiecti et passionem.

Ad 1.2 Ad aliam rationem, nego minorem; quamvis enim subiectum et sua definitio idem sint secundum rem, tamen illud quod subiectum dicit indistincte et confuse, definitio dicit distincte. Unde dicit Philosophus[c] I *Physicorum* quod definitio dividit in singularia, hoc est distincte indicat omnia principia definiti. Et ideo cum unumquodque magis cognoscatur

---

[a] Aristoteles, *Phys.*, IV, 4, 211 a 6–10   [b] Averroes, *Commentarium magnum in Physicam Aristotelis*, IV, 4, comm. 16 (1562, f. 133M)   [c] Aristoteles, *Phys.*, I, 1, 184 b 10–12

---

125 medium] subiectum M   134 solvatur] salvatur M   152 tamen] cum M   154 indicat] di *add. sed del.* M

in the subject. It is clear, then, as we have seen, that the essence of the subject is the middle in a demonstration of the strongest sort.

Further, the principle of demonstration is the middle term. I ask about that principle whether it is complex or simple. It cannot be complex, since that which is demonstrated and of which it is demonstrated is simple, and therefore the middle term is simple. If this is so, either it will be the essence of the attribute or the essence of the subject. It will not be the essence of the attribute, since this is demonstrable of the subject, therefore all that is left is that it be the essence of the subject. And this is what the Philosopher says in *Physics* IV, in the chapter on motion. For he says that such a definition of motion must be sought that provides from itself the essence of motion, and resolves the controversies, and which provides from itself the cause of all its attributes and properties. And the Commentator says there that perfect definitions of things are suited to provide the causes of all the accidents existing in those things. The middle term in a demonstration must thus be such that the cause of all the attributes can be drawn from it. Now this is the definition of the subject, since it is, through its formal principles, the cause of all the properties and attributes inhering in it, and this is to be understood in demonstration of the strongest sort. For it is still true that in demonstration not of the strongest sort the definition of the attribute can be the middle term of demonstrating one attribute of another, and this again can be the middle term for demonstrating yet another, while that one alone is demonstrated and does not demonstrate.

In response to the arguments.

Ad 1.1 In response to the first, I reply that it is not necessary that the middle term be of the same genus as the attribute so that it is of the same nature, but only of the same genus by that unity which exists between the cause and what is caused. But such unity exists between the essence of the subject and the attribute.

Ad 1.2 In response to the second, I deny the minor premise. For although the subject and its definition are in reality the same, nevertheless what the subject indicates indistinctly and confusedly the definition indicates distinctly. That is why the Philosopher says in *Physics* I that a definition divides into singulars—that is, it indicates distinctly every principle of

quando definite cognoscuntur sua principia quam quando confuse, ideo omnis definitio notior est definito, et ideo dicit Avicenna[a] quod definitio datur causa innotescendi.

⟨Questio 49⟩

Quia Philosophus[b] dicit quod cognoscentem quid est impossibile est ignorare si est, queritur utrum esse sit additum essentie.

1.1 Et videtur quod sic, quia Boethius[c] dicit super primum quod in omni quod est citra primum differt quod est et quo est; sed quod est est ipsa essentia et quo est esse; ergo etc.

1.2 Item, sicut vivere ad viventem, sic esse ad ens, quia vivere viventibus esse est;[d] sed vivere est additum viventi, quia vivere est movere secundum locum et sentire, secundum Philosophum[e] in *De anima*, sentire autem et movere secundum locum sunt addita viventi; ergo etc.

1.3 Item, hoc arguitur ratione Avicenne,[f] quia omne quod habet esse per essentiam suam ex se ipso determinatum est ad esse, et quod ex se ipso determinatum est ad esse, ex se determinatum est esse tale quod necesse est esse, ⟨unde⟩ est ipsum suum esse; et tale tantum est unum, scilicet causa prima; ergo in rebus omnibus aliis a primo esse erit aliud ab essentia.

1.4 Item, hoc arguitur ratione Thome[g] in tractatu suo *De esse et essentia*: ipsum esse non multiplicatur nisi per aliquod [additum] sibi additum, esse enim unum huiusmodi non est de se multiplicatum; et quia esse primi non est aliquod additum, ideo non est multiplicatum; si ergo esse in istis multiplicatur, hoc erit per aliquod additum.

1.5 Item, hoc arguitur ratione Alberti[h] in commento suo super librum *De causis*: omnis effectus primi recedit a simplicitate primi, et ideo in omni

---

[a] *Rectius* Aristoteles, *Top.*, VI, 1, 139 b 14–15 [b] Aristoteles, *AnPo*, II, 8, 93 a 20–23 [c] Boethius, *De trinitate* 2.92–94 *secundum sententiam Thomae*; cf. Thomas de Aquino, *In I Sententiarum*, d. 8, q. 5, a. 1 [d] Cf. Aristoteles, *De an.*, II, 4, 415 b 13 [e] Aristoteles, *De an.*, I, 1, 402 a 6–7 [f] Avicenna latinus, *De Philosophia prima vel scientia divina*, VIII, 3 (ed. Van Riet, pp. 395–97) [g] Thomas de Aquino, *De ente et essentia*, IV [h] Albertus Magnus, *De causis et processu universitatis a prima causa*, II, 1, 7 (ed. Fauser, p. 100)

---

3 boethius] philosophus M   7 est[1]] entibus M

what is defined. And therefore each is better cognized when its principles are cognized definitely than when they are cognized confusedly. Therefore every definition is better known than what is defined, and therefore Avicenna says that the definition is called the cause of becoming known.

Question 49.

Since the Philosopher says that it is impossible for one who cognizes what something is to be ignorant of whether it is, it is asked whether being (*esse*) is something added to essence.

1.1 It seems that it is, since Boethius says, concerning the first cause, that in everything other than the first cause, what something is and that by which something is differ, but what something is is its essence, and that by which something is is its being; therefore etc.

1.2 Furthermore, as living stands to a living thing, so being stands to an entity (*ens*), since, for living things, living is being. But living is added to a living thing, since living is locomotion and sensing, according to the Philosopher in *De Anima*, but sensing and locomotion are added to living thing. Therefore etc.

1.3 Furthermore, this is argued for through Avicenna's argument, since everything that has being through its essence is determined from itself to being, and what is determined from itself to being is determined to being in such a way that its being is necessary, hence it is its own being, and only one thing is like that, namely the first cause. Therefore in all things other than the first cause being will be other than essence.

1.4 Furthermore, Thomas argues for this in his treatise, *On Being and Essence*. Being is not multiplied except through something added to it, for one being of this sort is not of itself multiplied, and since the being of the first is not something added, therefore it is not multiplied. If, therefore, being in these others is multiplied, this will be through something added.

1.5 Furthermore, Albert argues for this in his commentary on *De causis*. Every effect of the first cause recedes from the simplicity of the first

effectu primi invenitur aliqua ratio compositionis, sed substantie separate a materia sunt effectus primi, ergo in eis est compositio; sed non est ibi compositio materie cum forma nec subiecti cum accidente, ergo in ipsis necessario reperietur compositio esse cum essentia; ergo multo fortius in istis inferioribus.

2.1 Oppositum patet per Philosophum[a] IV *Metaphysice*; dicit enim quod substantia uniuscuiusque est simpliciter una et simpliciter ens; et ens includit esse et habens esse, ergo substantia uniuscuiusque est essentialiter habens esse; ergo esse erit de essentia uniuscuiusque, quare non erit additum.

2.2 Item, Commentator[b] super IV *Metaphysice* dicit: si dicatur sic 'homo est', li 'est' potest predicare esse diminutum in anima vel esse in effectu. Si dicat esse diminutum in anima, tunc ⟨est⟩ problema de accidente, si esse in effectu, tunc est problema de genere. Sed in tali problemate [predicatione] predicatum et subiectum pertinent ad eandem essentiam. Ergo etc.

3 Istud est axioma de quo multi dubitant, secundum quod recitat Commentator[c] super IV *Metaphysice* opinionem Avicenne; dicit quod Avicenna dicit quod unumquodque causatorum sit ens per aliquod additum | essentie. Et duo erant ipsum moventia: unum, quod posuit ens esse de genere denominatorum, esse autem tale dicit intentionem subiecti et accidentis, unde cum ens dicit Philosophus esse, posuit ens sicut subiectum et esse sicut accidens subiecti. Item, dicebat quod ens et res imaginantur in anima tamquam due intentiones et non una, et ista duo significant aliud et aliud in omnibus linguis; sed non significarent aliud et aliud nisi unum adderet super aliud in significando, res autem, ut posuit, significat essentiam rei; et ideo posuit esse ut additum sibi et accidens essentie.

M 110va

Si Avicenna posuerit quod esse sit res addita essentie, videtur quod falsum posuit, quia si dicatur quod esse sit accidens reale additum, ergo erit ponere decem genera accidentium; sed hoc est impossibile; ergo etc. Probatio consequentie, quia si esse sit accidens, aut erit in genere qua-

---

[a] Aristoteles, *Met.*, IV, 2, 1003 b 16–33   [b] Averroes, *Commentarium magnum in Metaphysicam Aristotelis*, V, 7, comm. 14 (1562, f. 117G–H)   [c] Averroes, *Commentarium magnum in Metaphysicam Aristotelis*, IV, 2, comm. 3 (1562, f. 67B-I); V, 7, comm. 14 (1562, f. 117C–D)

---

34 li]   homo *add. sed del.* M   39 opinionem]   opinioni M   52 quia]   quod M

BOOK II   245

cause, and therefore in every effect of the first one finds some account of composition. But substances separated from matter are effects of the first cause; therefore there is composition in them. But there is no composition there of matter with form, nor of subject with accident, therefore in these there is necessarily found a composition of being with essence; therefore this is so all the more in those inferior to them.

2.1 The opposite is clear from the Philosopher in *Metaphysics* IV, for he says that the substance of each is one without qualification and a being without qualification, and an entity includes being and the having of being. Therefore the substance of each is essentially something that has being, and therefore being will be of the essence of each thing, and so it is not something added.

2.2 Furthermore, the Commentator on *Metaphysics* IV says that if one says 'a human being is', the 'is' can predicate a diminished being in the soul or being in actual effect. If it indicates diminished being in the soul, then it is a case of an accident; if it indicates actual being, then it is a case of a genus. But in such a case the predicate and subject belong to the same essence. Therefore etc.

3 Reply. This is an axiom concerning which many have some doubt, as the Commentator reports on *Metaphysics* IV when discussing the opinion of Avicenna. He says that Avicenna held that each thing that is caused is an entity through something added to its essence. Two claims moved Avicenna to this view. First, he held that *entity* is in the genus of denominators, whereas *being such* refers to the intention of subject and accident. Hence when the Philosopher says that an entity has being, he posits an entity as the subject and being as an accident of the subject. Further, he said that *entity* and *thing* are imagined in the soul as two intentions rather than one. These two signify differently in all languages, but they would not signify differently unless one of them were to add something to the other in signifying thing. His claim, however, is that *thing* signifies the essence of a thing, and so he posited *being* as something added to it and as an accident of its essence.

If Avicenna held that being is a thing added to the essence, it seems that he posited something false. For if he said that being is a real added accident, then one would need to posit ten genera of accidents. But this is impossible; therefore etc. The proof of the inference is that if being is

litatis aut quantitatis, et sic deinceps; sed de nullo istorum potest esse, quia omnia novem genera accidentium presupponunt esse, nihil autem se ipsum presupponit; quare erunt decem genera accidentium.

Quidam dicebant quod esse ⟨in effectu⟩ est in genere actionis. Et hoc est falsum, quia actioni est aliquod contrarium; sed esse in effectu non est aliquid contrarium; ergo etc. Item, actio presupponit aliquam qualitatem accidentalem; sed esse in effectu nullam presupponit, sed omnis qualitas accidentalis presupponit esse in effectu. Falsum est ergo dicere quod est in genere actionis.

Ea autem ex quibus fuit motus Avicenna non concludunt. Cum enim dicit quod ens est de genere denominatorum et cum dicit ⟨quod⟩ res et ens significant diversa, Commentator respondet dicens quod deceptus fuit, quia non distinxit inter nomina significantia diversas intentiones et eandem intentionem [modo tamen diverso]. Unde, sicut dicit, tres sunt nomina que convertuntur in suppositis, quedam significant diversas intentiones, ut homo et risibilis, et tamen convertuntur in suppositis, quedam eandem intentionem modo [tamen] eodem, ut tunica et vestis, quedam autem eandem intentionem, modo tamen diverso, ut unum et ens. Et ideo non est nugatio dicendo 'ens unum', et propter hoc unum verificatur per aliud, ut ens per unum.

Frater Thomas[a] reprobat Avicennam in hoc quod dicit quod esse sit aliquod additum sicut accidens; concedit enim quod sit quid additum, sed non sicut accidens, sed quod sit quid constitutum per principia essentialia rei cuius est esse.

Si intelligat sicut Avicenna intellexit, quod sit quid reale additum, male intelligit, quia esse esset quidam actus: aut ergo actus primus aut secundus. Non actus primus, quia actus primus non differt ab essentia rei; sequetur tunc quod esse non sit additum rei. Non secundus, quia talis actus est operatio ut procedens a forma; omnis autem talis actus secundus presupponit esse in effectu, nihil autem se ipsum presupponit; manifestum est ergo quod esse in effectu non est actus secundus. Si ergo dicatur quod esse in effectu sit quid reale additum preter istos duos actus, oportet ponere actum tertium in entibus, quod est contra Philosophum.

---

[a] Thomas de Aquino, *De potentia dei*, q. 5, a. 4, ad 3

---

53 sed] et M   54 quia] quod M   56 est[1]] esset M   ‖ actionis] accidentis *ante corr.* M
69 eandem intentionem] diversas intentiones M

an accident, it will be in the genus of either quality or quantity, and so on in order. But it cannot be in any of these, because all nine genera of accidents presuppose being. But nothing presupposes itself. Therefore there will be ten genera of accidents.

Now some have said that being in actual effect is in the genus of action. But this is false, since there is a contrary to action, whereas being in actual effect is not one of a pair of contraries, therefore etc. Also, action presupposes some accidental quality, whereas being in actual effect presupposes nothing, while every accidental quality presupposes being in actual effect. It is false, then, to say that it being is in the genus of action.

The claims that moved Avicenna are not successful. For when he says that *entity* is in the genus of denominators, and when he says that 'thing' and 'entity' signify differently, the Commentator responds by saying that Avicenna was deceived because he did not distinguish between names signifying different intentions and names signifying the same intention. Hence, as he says, there are three sorts of names that are interchangeable in what they refer to. Some signify different intentions, for instance, 'human being' and 'being able to laugh', and still are interchangeable in what they refer to. Others signify the same intention in the same way, like 'clothing' and 'garment'. Still others signify the same intention in different ways, like 'one' and 'entity'. And thus it is not trivial to speak of 'one entity', and this is why either is derivable from the other, for instance, being an entity from being one.

Brother Thomas argues against Avicenna's claim that being is something added as an accident. For he grants that it is something added—not as an accident, but as something constituted through the essential principles of the thing that has being.

If he thinks, as Avicenna thought, that being is something real added, then he is wrong, for being would then be a certain actuality, and so either first or second actuality. Not first actuality, since first actuality does not differ from the essence of the thing, and so then it would follow that being is not added to the thing. Nor is it second actuality, since such an actuality is an operation that proceeds from a form, but every such second actuality presupposes being in actual effect. Nothing, however, presupposes itself. Therefore it is obvious that being in actual effect is not second actuality. If, then, it is said that being in actual effect is something real added above these two actualities, then one has to postulate a third actuality in entities, which is contrary to the Philosopher.

Item, in re habente essentiam non invenimus nisi tria, scilicet essentiam et principia essentie et accidentia. Si ergo Thomas ponit quod esse in effectu nullum horum, ergo esse in effectu nihil erit.

Albertus[a] in commento suo super librum *De causis* ponit quod esse in effectu sit additum essentie, motus ex ista ratione, quia omne ens causatum habet esse a primo principio; nullum autem ens causatum habet suam essentiam ab alio, sed a se ipso; manifestum est igitur quod in quocumque ente causato esse erit aliud ab essentia cuius est esse.

Ista ratio non videtur concludere; necessarium enim est quod omne illud quod est posterius primo vadat in primum tamquam in suam causam. Cum ergo esse et essentia cuiuscumque entis causati sint posteriora primo principio, vadent in primum tamquam in suam causam, qua ergo ratione causata recipiunt esse a primo, eadem ratione et essentiam.

Omissis ergo istis opinionibus, videtur esse aliter dicendum. Ad cuius evidentiam tria sunt consideranda: primum est quod de nullo ente causato verum est dicere quod ipsum est suum esse; secundum est quod esse est aliquid additum causatis; et tertium quod esse non est res addita causatis, unde est additum et non est res addita.

Probatio primi est, quia in entibus causatis esse non est aliud quam quidam ordo vel respectus ad producens vel generans; unde aliquod causatum, puta homo, dicitur esse ens quia habet respectum ad suum generans per quod introductum est in entibus; ergo ⟨in⟩ causatis esse non est nisi ordo ad generans. Sed manifestum est quod nullum entium causatorum est ordo vel respectus. Ergo nullum ipsorum est ipsum suum esse.

Probatio secundi est, quia omne illud preter quod potest haberi | intellectus rei essentialis, et completus, illud est additum rei; sed preter esse potest haberi intellectus rei completus et essentialis, unde dicit Algazel[b] principio sue logice quod possumus intelligere quaternarium non intelligendo si est quaternarius; esse ergo est aliquid additum rei. Et confirmatur ratio, quia per eadem principia essentialia intelligo rem re existente et

M 110vb

---

[a] Albertus Magnus, *De causis et processu universitatis a prima causa*, I, I, 8 (ed. Fauser, p. 16)
[b] Al Ghazali, *Logica* (ed. Lohr, p. 247)

---

96 posteriora] causata M   105 unde] bene M   109 est ordo] *iter.* M

Further, in a thing having an essence only three things are found, namely the essence, the principles of the essence, and accidents. If, therefore, Thomas posits that being in actual effect is none of these, then being in actual effect will be nothing.

Albert, in his commentary on *De causis*, posits that being in actual effect is added to essence on the basis of this argument, that every caused entity has its being from the first principle. No caused entity, however, has its essence from another, but from itself. Therefore it is obvious that, in every caused entity, its being will be something other than the essence that has being.

This argument does not seem successful. For it is necessary that everything that is posterior to the first principle go back to the first as to its cause. Since, therefore, the being and essence of every caused entity are posterior to the first principle, they go back to the first as to their cause. Thus, for the same reason that entities that are caused receive their being from the first principle, so they receive their essence from the first principle.

Leaving aside these opinions, then, it seems we should respond otherwise. To make this clear there are three things to consider. The first is that there is no caused entity of which it is true to say that it is its own being. The second is that being is something added to entities that are caused. The third is that being is not a thing added to caused entities, and so it is something added but is not an added thing.

The proof of the first is that in caused entities being is nothing other than a certain order or relation to what produces or generates it. Accordingly, a caused entity, such as a human being, is said to be an entity because it has a relation to what generates it, through which it is introduced among entities. Therefore in caused entities being is nothing other than a certain order to what generates. But it is obvious that none of these caused entities is an order or relation. Therefore none of these is its own being.

Proof of the second point is that everything without which a complete and essential understanding of a thing can be had is something added to that thing. But a complete and essential understanding of a thing can be had without being. This is why Alghazali says in the beginning of his logic that we can understand a notebook without understanding whether a notebook exists. Therefore being is something added to a thing. And

non existente; facta ergo variatione in existentia rei, non oportet variationem fieri in intellectu essentiali rei; quare esse et non esse est additum essentie rei.

Probatio tertii, et hoc contingit declarare ratione Commentatoris IV *Metaphysice*.[a] Arguamus enim sic: homo est ens, aut ergo est ens per suam substantiam aut per aliquid additum sue substantie. Si sit ens per suam substantiam, propositum habetur; si per additum, quero de illo addito, quia oportet quod sit ens, aut ergo erit ens per suam substantiam aut per aliquid additum. Si per suam substantiam, eadem ratione fuit standum in primo; si per aliquid additum, erit processus in infinitum vel erit dare quod esse sit de essentia rei.

Hanc autem rationem Commentatoris aliqui superficialiter considerantes nituntur destruere. Cum ergo queritur utrum sit ens per suam substantiam aut per rem additam, dicunt quod per rem additam ut per esse. Et cum queritur de illo: illud erit ens? Dicunt quod falsum est, sed est illud quo aliud est ens, sicut ipsi ponunt exemplum: homo est albus per albedinem, et tamen albedo non est alba, sed albedo est illud quo homo est albus. Eodem modo dicunt quod esse non est ens, sed est illud quo aliud est ens.

Sed istud non valet. Accipiam enim ens quod est predicatum de omni causato et etiam de causa prima, quo ente nihil est latius. Loquendo de tali ente, ego quero utrum esse sit ens. Si sic, propositum habetur. Si non, cum omne quod est preter illud esse sit nihil, cum homo sit formaliter ens per tale esse, ergo homo et quodcumque aliud formaliter erit ens per nihil, quod est impossibile. Et iterum non diceret aliquis quod esse non esset ens nisi quia non est compositum ex esse et habente esse; sed non obstante ista compositione potest dici quod unum sit ens; ergo multo fortius esse erit ens preter hoc quod non sit compositum.

---

[a] Averroes, *Commentarium magnum in Metaphysicam Aristotelis*, IV, 4, comm. 14 (1562, ff. 80L–81A)

---

119 commentatoris] comm(entatoris) s' M   120 suam] *iter. sed corr.* M   128 nituntur] utuntur M

the argument is confirmed since I understand a thing through the same essential principles, whether or not the thing exists. So when a thing's existence is made to vary, it is not necessary to introduce a variation in the essential understanding of the thing. Hence being and non-being are added to a thing's essence.

Proof of the third point. One can make this clear through the argument of the Commentator on *Metaphysics* IV. Let us argue as follows: A human being is an entity; therefore it is an entity either through its substance or through something added to its substance. If it is an entity through its substance then we have what was proposed. If through something added then I ask about that added thing. Since it must be an entity, it is either an entity through its substance or through something added. If through its substance then for the same reason we ought to have stopped here initially. If through something added then there will be an infinite regress or it will have to be granted that being belongs to the thing's essence.

Now some, considering the matter superficially, try to destroy this argument of the Commentator's. For when it is asked whether it is an entity through its substance or though some added thing, they say that it is through some added thing, namely, through being. And when it is asked, regarding that added being, whether it will be an entity, they reply that it will not, and that instead it is that by which something else is an entity. As an example of this, they offer the way that a human being is white through whiteness, and yet the whiteness is not white, but rather the whiteness is that by which a human being is white. In the same way, they say that being is not an entity, but instead is that by which something else is an entity.

But this does not work. Take the entity that is predicated of every caused thing, and also of the first cause, the entity than which none is broader. Speaking of such an entity, I ask whether its being is an entity. If so, we have what was proposed. If not, then since everything that is outside that being is nothing, given that a human being is formally an entity through such being, it follows that the human being, and whatever else, will be formally a being through nothing, which is impossible. Moreover, no one would say that being is not an entity unless this is because it is not composed from being and something having being. But such composition notwithstanding, it can be said that what is one is an entity. Therefore being will all the more be an entity, aside from its not being composite.

Item, hoc ostenditur alia ratione: si homo sit ens per rem additam, aut erit illud additum de genere substantie aut accidentis. Non potest dici quod sit de genere substantie, quia ibi non sunt nisi tria, scilicet materia et forma et aggregatum, quorum nullum est hoc esse, et si sic, aliquod illorum non erit ens per aliquod ens. Item, non potest dici quod sit ens per aliquam rem additam que sit de genere accidentis, quia omne accidens presupponit esse in effectu. Et preter istas res non est alia res; quare homo non est ens per rem additam ipsi rei vel essentie in rebus causatis.

Apparet ergo quod esse est aliquod additum et non est res addita, ex quibus sequitur quod sit ratio addita; homo ergo et quodcumque causatum erit ens non per rem additam sed per rationem additam. Ista autem ratio addita aut erit absoluta aut respectiva. Non potest dici quod homo sit ens per rationem absolutam, quia ut sic magis homo haberet esse in anima quam in re extra, sed respectum habet ad propria accidentia, et ideo sequitur quod sit ens non per rationem aliquam absolutam. Erit ergo ens per rationem aliquam respectivam ad accidens aut [per] ad agens: non per rationem respectivam ad aliquod accidens, quia nullum accidens precedit esse in effectu, sed omne accidens necessario presupponit esse in effectu; quare sequitur necessario quod homo sit ens formaliter per respectum ad agens, illud enim solum precedit ipsum in natura. Unde si queratur a me per quid homo est ens formaliter, dico quod per hoc quod est terminus huius generationis; substantie autem separate sunt entia per hoc quod sunt termini huius factionis, sed ista sunt entia per hoc quod sunt termini huius factionis ab hoc agente et ab hac materia. Iste autem respectus est aliquid reale quia potest habere esse preter intellectum, sed tamen non est tale reale quod sit de genere accidentis, sed est in eodem genere cum re in quo est res cuius ipsum est. Et ita exponit Egidius[a] fratrem Thomam, ita quod esse sit additum, sed non sicut accidens reale sed sicut accidens rationis, et est idem quod est unumquodque illorum cuius est.

---

[a] Egidius Romanus, *De esse et essentia, de mensura angelorum, et de cognitione angelorum*, q. 9 (1503, f. 20v); *Theoremata de esse et essentia* (ed. Hocedez, th. XXII)

---

155 erit] ab *add. sed del.* M   157 sed] quia M   159 ad²] aliquid M

Further, this is shown by another argument. If a human being is an entity through some added thing, what is added will belong to the genus of either substance or accident. It cannot be said that it belongs to the genus of substance, since there are only three in this genus, namely matter and form and the aggregate, and none of them is this being. If this is so, then one of them will not be an entity through some entity. A human being also cannot be said to be an entity through some added thing that belongs to the genus of accident, since every accident presupposes being in actual effect, and apart from these ⟨accidental⟩ things there is no other thing. Therefore a human being is not a being through a thing that is added to the thing itself or its essence in caused things.

It is apparent, therefore, that being is something added, but that it is not an added thing. From this it follows that it is an added aspect (*ratio*). A human being, therefore, and any of the things that are caused, will be an entity not through any added thing, but through an added aspect. But this added aspect will be either absolute or relative. It cannot be said that a human being is an entity through an absolute aspect, since as such a human being would have being in the soul more than in external reality. But a human being has a relation to its proper accidents, and so it follows that it is not an entity through some absolute aspect. Therefore it will be an entity through some aspect that is relative to an accident or else to an agent. Not through an aspect that is relative to some accident, since no accident precedes being in actual effect, but rather every accident necessarily presupposes being in actual effect. Therefore it follows necessarily that a human being is an entity formally through its relation to an agent. For that alone precedes it in nature. Hence if someone asks me about that through which a human being is formally an entity, I reply that it is through its being the end result of this generation. Now separated substances are entities through their being the end results of a certain making, but human beings are entities through their being the end results of a certain making by this agent and this matter. And this relation is something real, since it can have being beyond the intellect. But still it is not something real of the sort that belongs to the genus of accident. Instead, it is in the same genus as the thing to which it belongs. This is how Giles explains brother Thomas, so that being is something added but not as a real accident. Instead it is an accident of aspect, and it is the same as each of these to which it belongs.

Et si quereret aliquis: cum relationes quedam sint que immediate fundantur super substantiam, falsum vide[n]tur quod omnes relationes fundantur super accidens. Dico quod possibile est aliquas relationes esse de genere substantie; manifestum est enim quod materia respectum habet ad formam. Iste igitur respectus aut inest materie per [rationem] substantiam materie aut per additum substantie eius. Si sic, habetur propositum, quod aliquis respectus immediate fundatur super substantiam. Si non, ergo materia habebit respectum et ordinem ad illud additum, aut igitur per substantiam suam, et sic habetur propositum | et eadem ratione standum fuit prius. Si per aliquid additum, materia adhuc habet respectum et ordinem ad illud, et sic vel procedetur in infinitum vel erit dare aliquem respectum et relationem immediate fundatam super substantiam. Licet tamen non sit res addita substantie rei, tamen poterit esse aliquod accidens rationis, quia non competit essentie homini secundum quod homo; homo enim secundum quod homo neque est neque non est, ut probat Avicenna.[a]

M 111ra

Ad rationes igitur dicendum, quia procedunt viis suis; que enim probant quod est aliquod additum, bene probant quod [non] est aliquid additum sicut dictum est. Dicendum tamen per ordinem ad ipsas.

Ad 1.1 Ad Boethium,[b] dicendum quod ipse intellexit per hoc quod ipsum quod est et quo est non sunt idem secundum rationem intelligendi, et verum est; si autem intelligat quod differant secundum rem, falso intelligit, sicut ostensum est.

Ad 1.2 Ad secundum, dicendum quod vivere vel dicit principium vite vel operationem consequentem ad viventem. Si autem intelligitur de principio vite, sic se habet ad vivens sicut esse ad ens; sicut enim vivere non est distinctum a vivente, similiter nec esse ab ente. Si autem intelligitur vivere quod est operatio viventis, dico quod non similiter se habet esse ad ens.

---

[a] Avicenna latinus, *De philosophia prima vel scientia divina*, V, 1 (ed. Van Riet, p. 227) [b] Boethius, *Quomodo substantiae*, a. II, VII, VIII (ed. Rand, pp. 40–42)

178–179 substantiam] *i. m.* M   197 vivere] verum est M

And if someone were to ask, 'since there are some relations that are grounded immediately in their substance, it seems false that all relations are grounded in an accident', I reply that it is possible for some relations to belong to the genus of substance. For it is obvious that matter has a relation to its form. Therefore this relation inheres either in matter through the substance of the matter or through something added to its substance. If the first then we have what was proposed, that some relation is immediately founded on a substance. If not, then matter will have a relation and be ordered to something added. If, then, this occurs through its substance then we have what was proposed, and for the same reason we ought to have stopped earlier. If this occurs through something added, then matter still has a relation and order to what was added. And so either we proceed infinitely or there will be some relation immediately grounded in the substance. But although it is not a thing that is added to the thing's substance, still there could be some accident of aspect. For it does not belong to the essence of a human being considered as a human being, since human being considered as human neither does nor does not exist, as Avicenna proves.

In response to the arguments it should be replied that each goes through in its own way. For the ones that prove that being is something added do succeed in proving that it is something added in the way described above. But we should reply to them in order.

Ad 1.1 In response to Boethius it should be replied that what he meant by this was that what something is and that by which something is are not the same according to the account by which they are understood, and this is true. But if he means that they differ as things, then this is false, as has been shown.

Ad 1.2 In response to the second it should be said 'living' refers either to the principle of life or to the operation that follows from living. If it is understood as referring to the principle of life, then it is related to the living thing just as being is related to an entity. For just as living is not distinct from a living thing, in the same way neither is being distinct from an entity. But if living is understood as the operation of something living, then I reply that it is not related to a living thing in the same way as being is related to an entity.

Ad 1.3 Ad tertium, dicendum quod 'ex' vel dicit circumstantiam cause efficientis, et sic maior est falsa; omne enim quod habet formaliter esse ex se determinatur ad esse ex se sicut ab efficiente. Si autem dicat circumstantiam cause formalis, sic vera est propositio; omne enim quod est formaliter ex se habet esse. Et accipiendo li 'ex' primo modo ∗∗∗

Ad 1.4 Ad quartum, 'esse non multiplicatur nisi per additum', dico quod ista ratio supponit quod ipsum esse quantum est de se sit unius rationis in omnibus entibus, quod tamen falsum est, immo est alia ratio entis secundum quod dicitur de substantia et secundum quod dicitur de accidentibus. Modo si istud esset verum, ipsum non multiplicaretur nisi per aliquod additum, nunc autem non est eiusdem rationis, et ideo potest multiplicari non per aliquid additum. Unde sicut substantia hominis ex se non per aliquod additum diversa est a substantia asini, similiter esse hominis ex se non per aliquod additum diversum est ab esse[ntia] asini.

Ad 1.5 Ad quintum, dicendum quod omnia citra primum recedunt a simplicitate primi, et ab actu primi, et accedunt ad potentiam. Et si per istum accessum et recessum ad potentiam intelligitur compositio, dicendum quod in omni citra primum est quedam compositio, et ex hoc dissolvitur ratio. Et verum est quod in istis substantiis separatis est aliqua compositio in hoc quod recedunt a simplicitate primi. Et cum dicitur 'sed in ipsis non est compositio materie cum forma, nec accidentis cum subiecto, quare erit compositio essentie cum esse', dico quod est fallacia consequentis, immo preter illam compositionem dabo aliam, ut potentie et actus, vel compositionem rationis vel actus intelligendi. Intelligunt enim per rationem aliam intelligendi a substantia earum; nulla enim earum est suum intelligere. Ipsum autem primum non intelligit per rationem aliam intelligendi a substantia eius, et ideo in ipsis substantiis separatis aliis a primo est talis compositio, non autem est ibi compositio essentie cum esse.

---

207 ∗∗∗] *spatium vacans fere 23 litterarum* M   213 ideo] non *add. sed exp.* M   218 primi²] ponitur M   219–220 dicendum ... compositio] *iter. sed corr.* M   221 est²] substantia *add. sed exp.* M   222 ipsis] ipso M   225 preter] per *ante corr.* M

Ad 1.3 In response to the third it should be replied that if 'from' refers to the circumstance of an efficient cause, then the major premise is false, for everything that has being formally from itself is determined to being from itself as from an efficient cause. But if 'from' refers to the circumstance of a formal cause, then the proposition is true, since everything that exists has being formally from itself.

Ad 1.4 In response to the fourth, that being is not multiplied except through something added, I reply that this argument assumes that being as it is in itself has the same account in all beings, which is false. Instead, being has a different account as it is said of substance and as it is said of accidents. Now, if it were true that it has the same account in all beings, then it would be multiplied only through something added. But in fact it does not have the same account, and therefore it can be multiplied without anything's being added. So, for instance, the substance of human being, of itself and not through anything added, is different from the substance of a donkey. In the same way, the being of a human being, of itself and not through anything added, is different from the being of a donkey.

Ad 1.5 In response to the fifth argument it should be replied that all things aside from the first cause recede from the simplicity and actuality of that first cause and draw near to potentiality. If this drawing near to and receding from potentiality is understood as composition, then it is to be replied that in everything other than the first cause there is a certain composition. In this way the argument is dissolved. And it is true that in these separated substances there is some composition in their recession from the simplicity of the first cause. And when it is said, 'but in these cases there is no composition of matter with form, nor of accident with subject, therefore there will be a composition of essence with being', I reply that this is the fallacy of the consequent. For in place of that composition, I offer another, such as that of potentiality with actuality, or the composition of aspect or act of understanding. For the separate substances understand through an account of understanding that is distinct from their substance, since none of them is its very understanding. The first cause, in contrast, does not understand through an account of understanding that is distinct from its substance. Therefore in these separate substances that are distinct from the first substance there is such a composition, but it is not the composition of essence with being.

⟨Questio 50⟩

Consequenter queritur utrum alicuius eiusdem sit definitio et demonstratio.

1.1 Et videtur quod sic, quia passionis est demonstratio; sed passionis est definitio; ergo eiusdem est definitio et demonstratio.

1.2 Item, cuiuscumque est demonstratio eius est principium demonstrationis; ergo cuiuscumque est demonstratio eius est definitio.

2.1 In oppositum est Philosophus.[a]

2.2 Et arguitur ratione quia demonstratio per se est eius quod est ens per accidens, quia demonstratio est conclusionis in qua demonstratur accidens inherere subiecto, et ita demonstratio est complexi; sed definitio non est entis secundum accidens nec complexi, sed solius entis per se et incomplexi; ergo etc.

3 Ad hoc dicendum quod nullius eiusdem cuius est demonstratio per se et primo est definitio per se. Et hoc apparet, quia definitio est primo eius quid est, et demonstratio per se est eius quia est; sed quid est et quia est aliud et aliud; quare nullius unius est primo definitio et demonstratio.

Item, demonstratio per se et primo est ipsius conclusionis in qua aliqua passio predicatur de subiecto; sed talis non est definitio per Aristotelem VII *Metaphysice*;[b] quare etc. Ex his apparet quod nullius eiusdem est definitio et demonstratio. Et hoc convertitur, cum sit negativa universalis, quare demonstratio Aristotelis plana est. Quamvis autem nullius per se et primo sit demonstratio et definitio, eiusdem tamen per accidens potest esse definitio et demonstratio, quia demonstratio est ipsius complexi per se et partium conclusionis est definitio, ideo eiusdem per accidens est definitio et demonstratio.

---

[a] Aristoteles, *AnPo*, II, 3, 90 a 35–91 a 11 [b] Aristoteles, *Met.*, VII, 12, 1037 b 9–27

9 quia] conclusio est demonstrationis *add. sed exp.* M   21 autem] enim M

Question 50.

Next it is asked whether there is ever both a definition and a demonstration of the same thing.

1.1 It seems that there is, since there is demonstration of an attribute, but there is also a definition of an attribute, and therefore there is definition and demonstration of the same thing.

1.2 . Furthermore, there is a principle of demonstration for whatever has a demonstration. Therefore, for whatever has a demonstration, there is a definition.

2.1 The Philosopher takes the opposite view.

2.2 This is also argued for by reason since a demonstration *per se* is of what is an accidental entity, since a demonstration is of a conclusion in which the accident is demonstrated to inhere in the subject, and so a demonstration is of something complex. But a definition is not of an entity with respect to an accident, nor of a complex, but only of a *per se* and simple entity; therefore etc.

3 It should be replied to this that there is no demonstration *per se* and primarily of anything of which there is also a definition *per se*. And this is apparent since definition is primarily of what-it-is, and demonstration *per se* is a demonstration that-it-is-so. But what-it-is and that-it-is-so are different. Therefore there is nothing of which there is both a definition and a demonstration primarily.

Further, a demonstration *per se* and primarily is of a conclusion in which an attribute is predicated of a subject. But there is no such definition, according to Aristotle in *Metaphysics* VII. Therefore etc. From these it is apparent that there is no one thing of which there is both a definition and demonstration. And this is convertible, since it is a negative universal, so the demonstration of Aristotle is plain. But even though there is nothing of which there is both a demonstration and a definition *per se* and primarily, still, there can be a definition and a demonstration accidentally of the same thing, for a demonstration *per se* is of a complex, and there is a definition of the parts of the conclusion. Therefore there is a definition and a demonstration of the same thing, accidentally.

Ad 1.1 Ad primam rationem, dicendum quod demonstratio non est solius passionis primo, sed demonstratio dicit causam inherentie predicati sive passionis ad subiectum, et ex consequenti est demonstratio passionis cuius est per se definitio.

Ad 1.2 Ad aliud, dicendum quod definitio bene est principium sed differt secundum quod est principium et secundum quod est principium inherentie passionis ad subiectum, sicut differt quid et propter quid. Et cum ipsa definitio diversimode sic accipitur, potest assignari fallacia accidentis in argumento, quia variatur medium.

⟨Questio 51⟩

Queritur utrum solius substantie sit definitio.

1.1 Et quod sic videtur Aristotelem[a] velle dicere in VII *Metaphysice*.

1.2 Et arguitur ratione, quia | definitio est solius entis per se; sed sola substantia est ens per se; quare etc.

M 111rb

2 Oppositum arguitur: omne quod habet quidditatem et genus et differentias potest definiri; sed talia sunt accidentia, que non sunt substantie; ergo etc.

3 Ad hoc dicendum quod ipsius substantie primo est definitio, et ex consequenti accidentium, sicut vult Philosophus[b] VII *Metaphysice*. Et ratio huius est, quia definitio dicit quid est rei, sed substantie primo habent quod quid est, et ideo substantiarum primo est definitio.

Item, quod quid est substantie ex alio non dependet, sed omnia alia accidentia dependent ad substantiam. Et ideo solius substantie est definitio primo, et aliorum ex consequenti et per habitudinem ad substantiam, quia accidentia dependent a substantia, et ideo accipiunt substantiam in definitione sua. Unde dico quod accidentium est definitio ex consequenti, quia accidentia habent quod quid est cum diminutione et secundum quid et per habitudinem ad substantiam, quia accidentia non sunt entia nisi quia entis, per Philosophum[c] VII *Metaphysice*, quare etc. Et hoc intelligit Aristoteles dicens quod definitio est substantie primo, et eius non est demonstratio.

[a] Aristoteles, *Met.*, VII, 5, 1031 a 1–3   [b] Aristoteles, *Met.*, VII, 5, 1031 a 1–3   [c] Aristoteles, *Met.*, VII, 5, 1031 a 1-3

26 solius] rationis *add. sed exp.* M   30 principium] inherentie *add. sed exp.* M   ‖ sed] et M

Ad 1.1 In response to the first argument, it should be held that a demonstration primarily is not only of the attribute, but it indicates the cause of the inherence of the predicate or attribute in the subject. Consequently, there is a demonstration of the attribute of which there is a definition *per se*.

Ad 1.2 In reply to the second argument it should be held that a definition certainly is a principle, but it differs as it is a principle and as it is a principle of the inherence of the attribute in the subject, just as what-it-is and because-of-what-it-is differ. And when the definition is thus taken in different ways, one can assign the fallacy of accident to the argument, for the middle term varies.

Question 51.

Next it is asked whether definitions are only of substances.

1.1 It seems that Aristotle holds that this is so, in *Metaphysics* VII.

1.2 This is also argued for by reason, because a definition is only of *per se* beings, but a substance alone is a being *per se*, therefore etc.

2 On the other hand, it is argued that everything that has quiddity, genus, and difference can be defined, but accidents are such, and they are not substances, therefore etc.

3 It should be replied that a definition is primarily of a substance, and of its accidents in consequence, as the Philosopher holds in *Metaphysics* VII. And his argument is that since a definition indicates the essence of a thing, but substances have their essences primarily, so too definitions are primarily of substances.

Further, the essence of a substance does not depend on anything else, but every accident depends on its substance. And so definitions are primarily of substances alone, and of others only in consequence, through their relation to their substance. For accidents depend on substances, and so they take substance into their definitions. Hence I maintain that definitions are of accidents in consequence, for accidents have an essence to a lesser degree and in a certain respect, through their relation to their substance. For accidents are beings only because they are of a being, according to the Philosopher in *Metaphysics* VII, therefore etc. And this is what Aristotle means, when he says that definitions are primarily of substances, of which there is no demonstration.

Ad 1.1 et 1.2 Et per hoc patet ad rationes, quia procedunt viis suis.

⟨Questio 52⟩

Consequenter queritur utrum quod quid est possit demonstrari.

1.1 Et arguitur quod sic, quia conclusionis demonstrationis est demonstratio; sed quod quid est aliquando concluditur in demonstratione, per Philosophum hic et in primo huius; quare etc.

1.2 ⟨...⟩

2 ⟨...⟩

3 Ad hoc est intelligendum quod duplex est quod quid est, quia quoddam est quod quid est subiecti et quoddam passionis. Unde quod quid est passionis potest demonstrari. Sed quod quid est subiecti dupliciter est, loquendo modo logico, quia quoddam est quod quid est quod est idem cum essentia subiecti, immo hoc est illud ipsum, et tale quod quid est non potest demonstrari. Aliud autem est quod quid est quod datur per aliquam causam extrinsecam, sicut per agens vel finem, et tale quod quid est est incompletum, cum non indicet essentiam rei perfecte, et tale quod quid est potest demonstrari, ut quod quid est quod datur per causam finalem potest demonstrari per causam agentem et e converso. Et hoc determinat Aristoteles[a] in littera.

Ad 1.1 Unde prima ratio bene probat quod aliquod quod quid est potest demonstrari, et hoc est verum de quod quid ⟨est⟩ passionis.

Ad 1.2 Ad secundum, dicendum quod sic arguendo: 'omne risibile est animal rationale etc.', bonus est syllogismus, ⟨sed⟩ nec est sophisticus nec demonstrativus; plures enim sunt modi syllogismorum quam hii, scilicet syllogismus conversivus et circularis etc.

---

[a] Aristoteles, *AnPo*, II, 8, 93 a 16–b 14

---

11 immo] in modo M   13 aliquam] aliam M   14 incompletum] incomplexum M
22 hii] habentur M

Ad 1.1 and 1.2 The response to the arguments is clear from this, for each goes through in its own way.

Question 52.

Next it is asked whether an essence (*quod quid est*) can be demonstrated.

1.1 It is argued that it can, for a demonstration is of the conclusion of a demonstration, but an essence is sometimes the conclusion of a demonstration according to the philosopher here and in *Posterior Analytics* I; therefore etc.[16]

1.2 ⟨...⟩

2 ⟨...⟩

3 In reply to this it should be understood that there are two kinds of essences. One is the essence of a subject and the other is the essence of an attribute. The essence of an attribute can be demonstrated. But the essence of a subject comes in two kinds, speaking logically. There is the essence (*quod quid est*) that is the same as what is essential (*essentia*) to the subject, and this is the essence itself, and as such it cannot be demonstrated. The other is the essence that is given through an extrinsic cause, for instance through an agent or end, and such an essence is incomplete, since it does not perfectly indicate what is essential to the thing. Such an essence *can* be demonstrated: for instance, the essence given through a final cause can be demonstrated through the agent cause, and conversely. This is how Aristotle resolves the issue in the text.

Ad 1.1 Hence the first argument proves well enough that some essence can be demonstrated, and this is true of the essence of an attribute.

Ad 1.2 In reply to the second, when someone argues that 'Everything that can laugh is a rational animal, etc.', the syllogism is a good one, but it is neither sophistical nor demonstrative. For there are more kinds of syllogism than these, such as convertible and circular syllogisms, and so on.

---

[16] A second initial argument, to which the text responds below, has been omitted at this point in the text. The customary argument(s) to the contrary are also missing.

Et si aliquis arguat quod ad destructionem unius speciei non sequitur destructio generis, dicendum quod syllogismus non se habet ad syllogismum sophisticum et dialecticum et demonstrativum sicut genus ad species, quia species addit novam formam supra genus. Sed syllogismus aut dialecticus vel demonstrativus vel sophisticus non addunt novum modum arguendi supra rationem syllogismi, sed materiam solum. Unde syllogismus simpliciter se habet ad hos syllogismos particulares sicut una forma vel species ad diversas materias, ut circulus se habet ad circulum eneum vel cupreum.

Et si quis arguat quod, cum ratio syllogismi non salvatur nisi in materia probabili vel sophistica vel necessaria, et si in materia sophistica, est syllogismus sophisticus, si in materia probabili, est syllogismus dialecticus, et si in materia necessaria, est syllogismus demonstrativus, et cum iste syllogismus 'omne risibile est animal rationale mortale; omnis homo est risibilis; ergo omnis homo est animal rationale mortale' sit in materia necessaria, sequitur quod iste syllogismus sit demonstrativus, dicendum quod non sequitur nisi haberet omnes conditiones demonstrationis perfecte. Unde cum non habeat omnes conditiones, quia non procedit ex causis conclusionis, ideo non est demonstratio, sed magis est petitio principii, cum premisse non probent conclusionem.

⟨Questio 53⟩

Consequenter queritur utrum quod quid est possit aliqualiter probari per syllogismum, et non dico demonstrari.

1 Et arguitur quod sic per Aristotelem,[a] qui ponit considerationes ad construendum quod quid est et definitionem de aliquo, quod patet ex VI *Topicorum*; quare etc.

2 Oppositum dicit Aristoteles hic.

3 Dicendum quod quod quid est simpliciter non contingit probari per aliquem syllogismum, quia ubi est petitio principii, ibi non est probatio; sed in omni syllogismo per quem concluditur quod quid est est petitio principii; ideo etc. Maior patet per Aristotelem hic. Ad aliquem tamen

[a] Aristoteles, *Top.*, VI, 1, 139 a 25

27 sed] et M   34 probabili] dialectica M

BOOK II   265

And if someone were to argue that the destruction of the genus does not follow from the destruction of one species, it should be replied that syllogism is not related to sophistical syllogism and dialectical syllogism and demonstrative syllogism as genus to species, for the species adds a new form over and above the genus. But a dialectical or demonstrative or sophistical syllogism does not add a new way of arguing to the account of being a syllogism, but adds only matter. Hence syllogism, without qualification, is related to those particular syllogisms as one form or species to different matters, just as circle is related to a bronze circle or a copper circle.

Now someone might argue that the account of syllogism is maintained only in plausible or sophistical or necessary matters. If in a sophistical matter then it is a sophistical syllogism; if in a plausible matter then it is a dialectical syllogism; and if in a necessary matter then it is a demonstrative syllogism. But this syllogism—'Everything that can laugh is a rational mortal animal, every human being is something that can laugh, therefore every human being is a rational mortal animal'—is in necessary matter. Therefore it follows that this syllogism is demonstrative. To this argument it should be replied that the conclusion follows only if all the conditions for a completed demonstration are satisfied. So since the syllogism does not satisfy all those conditions, given that it does not proceed from the causes of the conclusion, it is not a demonstration, but is rather a begging of the question, since the premises do not prove the conclusion.

Question 53.

Next it is asked whether an essence can in any way be proved through a syllogism, and I do not mean 'demonstrated'.

1 It is argued that it can from Aristotle, who offers consideration regarding the construction of the essence and the definition of a thing, as is clear from *Topics* VI. Therefore etc.

2 On the other hand is what Aristotle says here.

3 It should be replied that the essence without qualification is not able to be proved through any syllogism, for where there is a begging of the question there is no proof, but in every syllogism that has the essence as its conclusion the question is begged. Therefore etc. The major premise is obvious from Aristotle here. Even so, it is possible to prove the essence

contingit probare quod quid est de aliquo per probabilia, et sic facit Aristoteles in *Topicis*, unde nihil prohibet petitionem principii simpliciter esse probationem quoad aliquem.

Ad 1.1 Et per hoc patet ad rationem.

⟨Questio 54⟩

Consequenter queritur utrum in syllogismo concludente quod quid est sub ista reduplicatione 'in eo quod quid est' oporteat recipere istam reduplicationem in maiori et in minori propositione.

1 Et videtur quod non, quia sicut se habet quod quid est in premissis ad concludendum quod quid est in conclusione, sic se habet necessarium in premissis ad concludendum necessarium in conclusione; sed ad concludendum conclusionem de necessario non oportet utramque premissarum esse de necessario, sed sufficit quod maior sit de necessario; ergo etc.

2 Oppositum dicit Aristoteles.[a]

3 Dicendum quod quod quid est est predicatum convertibile cum subiecto, et medium per quo⟨d⟩ concluditur quod quid est de eo cuius est quod quid est est essentiale et proportionale extremis, et ideo oportet medium accipi sub ista reduplicatione '⟨in eo⟩ quod quid est' ad denotandum essentialem comparationem medii ad extrema. Unde si '⟨in eo⟩ quod quid est' adderetur uni extremo et non alteri, tunc significaretur quod comparatio medii ad unum extremum esset essentialis et ad aliud non, | et ideo si sic arguatur: 'omne B est A in eo quod quid est; omne C est B', non tenetur 'ergo omne C est A in eo quod quid est', quia minor non denotatur [f] esse necessaria, quamvis contingat eam esse necessariam.

M 111va

Ad 1 Ad argumentum in oppositum, dicendum quod ad concludendum conclusionem de necessario sufficit unam premissam esse de necessario, ad concludendum tamen quo⟨d⟩ quid est de aliquo non sufficit accipe-

---

[a] Aristoteles, *AnPo*, II, 4, 91 a 19–25

---

6 concludendum¹] quod quid est *add. sed exp.* M ‖ necessarium] *s. l.* M   12 essentiale] substantiale M

of something through plausible assumptions, when arguing for a specific audience (*ad aliquem*), and this is what Aristotle does in the *Topics*. Therefore nothing prevents an argument that begs the question without qualification from being a proof with respect to a specific audience.

Ad 1.1 And through this the reply to the argument is obvious.

Question 54.

Next it is asked whether, in a syllogism whose conclusion is an essence under the reduplication 'with respect to its essence', this reduplication must appear in the major and minor premises.

1 It seems that it need not, since just as an essence in the premises is related to a conclusion about that essence in the conclusion, so is necessity in the premises related to a conclusion of necessity in the conclusion. But, to reach a conclusion concerning something necessary it is not necessary for both premises to make a claim about necessity. Rather, it suffices that the major makes a claim about necessity. Therefore etc.

2 On the other hand is what Aristotle says.

3 It should be said that the essence is a predicate convertible with the subject, and the middle term through which a conclusion is reached about the essence of that of which it is the essence is essential and proportionate to the extremes. Therefore the middle term must appear under the reduplication 'with respect to its essence', to denote the essential relation of the middle term to the extremes. Hence, if 'with respect to its essence' were added to one extreme and not the other, then it would signify that the relation of the middle term to one extreme is essential, and to the other not. And so if it were argued thus: 'Every B is A in its essence, every C is B', it would not hold that 'therefore every C is A in its essence', because the minor is not denoted as being necessary even though it happens that it is necessary.

Ad 1 In response to the argument for the other side, it should be replied that, to reach a conclusion asserting necessity it is sufficient for one premise to assert necessity, but to reach a conclusion concerning the essence of something, it is not enough to rely on one premise with the

re unam premissam cum ista reduplicatione 'in eo quod quid est', sed oportet utramque sic sumi. Cuius ratio est, quia plus exigitur ad concludendum quod quid est de aliquo quam ad concludendum conclusionem; sufficit enim quod conclusio vera sit, et quod non possit esse falsa; ad concludendum autem quod quid est de aliquo ista requiruntur, et etiam plus: quod ipsum sit essentiale subiecto et etiam cum ipso convertibile.

⟨Questio 55⟩

Consequenter queritur utrum quod quid est possit ostendi via divisiva.

1 Et arguitur quod sic, quia omnem viam contingit quod quid est ostendere que omnia predicata essentialia accipit; sed talis est via divisiva; quare etc. Et hoc est quod videtur Aristotelem[a] velle in VII *Metaphysice*, ubi docet investigare quod quid est via divisiva; ergo etc.

2 Oppositum declarat hic Aristoteles.[b]

3 Dicendum quod quamvis quod quid est possit concludi per viam divisivam de aliquo, non tamen concluditur sub ista intentione quod quid est. Unde si sic arguatur: 'omne animal est rationale vel irrationale, et homo non est irrationale', ex his sequitur quod homo sit rationalis, supposito quod natura tota evacuetur per viam divisivam in propositione prima, non tamen rationale probatur de homine. Quamvis tota natura animalis per rationale et irrationale evacuetur, cum dicitur 'homo est rationalis', petitur; nam eque notum est quod unum contrariorum alicui inheret sicut remotio alterius contrarii ab eodem, et ita via divisiva non probat. Si tamen probaret, hoc non esset sufficienter ⟨in⟩ ostendendo quod quid est sub ratione eius quod quid est. Per predictam ergo rationem divisivam ostenditur quod homo sit rationalis; sed tamen quod homo sit rationalis essentialiter non ostenditur; quare talis via non ostendit quod quid est sub propria ratione eius.

---

[a] Aristoteles, *Met.*, VII, 12, 1037 b 29–1038 a 35   [b] Aristoteles, *AnPo*, II, 5, 91 b 12–92 a 5

7 quamvis] *s. l.* M   12 non] hoc M   14 quod] per M   18 rationalis²] rationale M

reduplication 'with respect to its essence', but one needs both of them to be taken like that. The reason for this is that more is needed for a conclusion about the essence of a thing than is needed to reach a conclusion. For the latter, it is enough that the conclusion be true, and that it cannot be false. But to reach a conclusion about something's essence these two things are required and more as well, namely, that it is essential to the subject and convertible with it.

Question 55.

Next it is asked whether the essence can be shown by the way of division.

1 It is argued that it can, since it is possible to show an essence in any way which takes up all the essential predicates ⟨underneath some genus⟩, but the way of division is such, therefore etc. And this is what Aristotle seems to hold in *Metaphysics* VII, where he teaches how to investigate an essence by the way of division. Therefore etc.

2 The opposite view is set out by Aristotle here.

3 It should be said that although one can reach a conclusion about an essence through the way of division, one cannot reach that conclusion when conceived of (*sub intentione*) as an essence. So if it is argued, 'Every animal is rational or nonrational, and human being is not nonrational', then from this it follows that a human being is rational, supposing that the whole nature is exhausted by way of the division in the first proposition, but there is no proof that *rational* belongs to human being. Even if the whole nature of being an animal is exhausted by the rational and the nonrational, when it is said that a human being is rational the question is begged, for it is just as well known that one of the contraries inheres in the subject as it is known that the other contrary is to be removed from that same subject. Thus the way of division does not prove anything. And even if it did prove this, this would not be sufficient to show the essence under its proper account. Therefore, the argument from division given above shows that a human being is rational, but it does not show that a human being is essentially rational. Therefore the way of division does not show an essence under its proper account.

Ad 1 Ad argumentum in contrarium, dicendum quod via divisiva accipiens omnia predicata essentialia sufficienter bene potest ostendere quod quid est alicuius de illo cuius est per remotionem aliorum diversorum essentialium. Illa tamen via non ostendit quo⟨d⟩ quid est inesse illi cuius est sub propria ratione eius quod quid est, et hoc vult Aristoteles hic et in VII *Metaphysice*. Aristoteles tamen hic ostendit quod quod quid est non potest demonstrari de aliquo sub propria eius ratione, docens accipere quod quid est cum reduplicatione in utraque premissarum. Vel potest dici: quod quid est simpliciter, quamvis non possit ostendi ⟨simpliciter, potest tamen ostendi⟩ ad aliquem, sicut prius dictum est.

⟨Questio 56⟩

Queritur utrum contingat ostendere quod quid est per definitionem eius.

1.1 Et arguitur quod sic, quia nulla est firmior ostensio quam illa que est per definitionem; cum igitur contingat quod quid est concludere esse quod quid est alicuius de aliquo per eius definitionem, manifestum videtur quod talis ostensio sit bona.

1.2 Item, omne quod predicatur de aliquo in quid et convertibiliter inest sibi per suum quod quid est; sed animal rationale predicatur de homine in quid et convertibiliter, ergo predicatum huius propositionis inest subiecto per suum quod quid est; quare etc.

2 ⟨...⟩

3 Dicendum quod quod quid est non potest demonstrari per definitionem, quia probatio videtur petere, cum definitio et definitum convertantur, et ita definitum non potest probari per definitionem, et hoc demonstrative. Quod quid est tamen potest ostendi dialectice per eius definitionem, sicut apparet ex II *Topicorum*,[a] sed talis ostensio non est simpliciter demonstratio sed secundum quid solum.

---

[a] Aristoteles, *Top.*, VII, 3, 133 a 11–24

Ad 1 In response to the argument to the contrary, it should be said that the way of division, since it takes up every essential predicate, can well enough show the essence of something whose essence it is, through the removal of the various other essential predicates. But this way does not show that the essence inheres in that whose essence it is under the account proper to the essence, and this is what Aristotle holds in *Metaphysics* VII. Here, however, Aristotle shows that an essence cannot be demonstrated of anything under its proper account, teaching that we should take the essence with reduplication in both premises. Alternatively, one can say of the essence without qualification that, although it cannot be shown without qualification, it can be shown to a specific audience, as was said before ⟨in q. 53⟩.

Question 56.
It is asked whether an essence can be shown through its definition.

1.1 It is argued that it can, since no way of showing something is stronger than through definition. Since, then, one can conclude something is the essence of something through its definition, it seems obvious that such a showing is a good one.

1.2 Furthermore, everything that is predicated of something substantially (*in quid*) and convertibly inheres in it through its essence. But rational animal is predicated of human being substantially and convertibly; therefore the predicate of this proposition inheres in the subject through its essence; therefore etc.

2 ⟨...⟩[17]

3 It should be replied that an essence cannot be demonstrated through its definition, for a proof seems to beg the question when the definition and what is defined are convertible. Thus what is defined cannot be proved, demonstratively, through its definition. Now an essence can be shown dialectically through its definition, as is apparent from *Topics* II, but such a showing is not a demonstration without qualification, but only in a certain respect.

---

[17] The manuscript provides no argument *in oppositum*.

Ad 1.1 Et per hoc patet ad rationes: nisi talis ostensio peteret; nunc autem petit simpliciter et secundum rem, quamvis alicui sufficiat talis ostensio pro demonstratione.

Ad 1.2 Et per hoc patet ad aliud.

⟨Questio 57⟩

Queritur utrum quod quid est unius contrarii possit ostendi per definitionem alterius contrarii.

1.1 Et arguitur quod sic, quia si sic dicatur: que maxime differunt contraria significant, ita unum contrariorum per aliud potest ostendi.

1.2 Item, contrariorum contraria sunt quod quid est, cuius ratio est, quia contrarietas est acceptio alicuius oppositionis secundum formam; et ita si unum contrariorum definiatur, opposita definitio ostendit quod quid est alterius contrarii; quare etc.

2 Oppositum docet Aristoteles.[a]

3 Dicendum quod unum contrari⟨or⟩um non potest ostendi per aliud simpliciter, immo simpliciter est petitio, quia unum contrari⟨or⟩um non magis est notum de aliquo quam alterum, et ideo unum non potest per aliud ostendi simpliciter, secundum quid tamen unum contrari⟨or⟩um habet ostendi per aliud.

Et per hoc patet ad rationes; procedunt enim viis suis.

⟨Questio 58⟩

Queritur utrum cognoscentem quid est oporteat cognoscere si est.

1.1 Et arguitur quod sic, quia quod non est non contingit scire; ergo quod contingit scire, ipsum est; quare etc.

---

[a] Aristoteles, *AnPo*, II, 6, 92 a 20–34; *Top.*, VII, 3, 153 a 26–b24; VI, 9, 147 a 29–b 25

---

3 maxime] bene M   6 oppositionis] propositionis M

Ad 1.1 From this it is clear what to say to the arguments. Such a showing would be good, unless it begs the question. But in fact it does beg the question, both without qualification and in reality, even if such a showing suffices as a demonstration for some.

Ad 1.2 From this the reply to the other is obvious.

Question 57.

It is asked whether the essence of one contrary can be shown through the definition of the other contrary.

1.1 It is argued that it can. For if it is said that things that differ maximally signify contraries, then one contrary can be shown through the other.

1.2 Furthermore, the essences of contraries are contraries, because contrariety is the recognition of some opposition with respect to its form. So if one contrary is defined then the opposed definition shows the essence of the other contrary; therefore etc.

2 Aristotle teaches the opposite view.

3 It should be replied that one contrary cannot be shown through the other without qualification. Rather, without qualification it begs the question, for one contrary is not better known in any subject matter than is the other. And therefore one cannot be shown without qualification through the other, even if one contrary is able to be shown through the other in a certain respect.

And from this it is clear how to reply to the arguments, for each goes through in its own way.

Question 58.

It is asked whether one who cognizes what-it-is must cognize if-it-is.

1.1 It is argued that he must, since it is not possible to know what does not exist, therefore what one can know exists; therefore etc.

1.2 Item, sicut propter quid se habet ad quia est, sic quid est se habet ad si est, per Aristotelem[a] in hoc secundo; sed si non contingit scire quia est, non contingit scire propter quid est; ergo etc. |

2 Oppositum dicit Aristoteles;[b] dicit enim quod contingit scire quid est triangulus et tamen dubitare si est; quare etc.

3 Ad hoc est intelligendum quod questio si est duo potest querere: potest enim querere si aliquid est absolute vel si est aliquid in effectu. Si questio querat primo modo, tunc questio quid est presupponit questionem si est, quia illud quod primo ponitur in quod quid est alicuius est natura entis, et ideo ratio entis primo occurrit intellectui de unoquoque. Si autem questio si est querat secundo modo, ⟨tunc questio quid est non presupponit questionem si est⟩, cum esse in effectu non sit de ratione entium causatorum, per Aristotelem[c] hic consequenter qui dicit quod si est sive esse potest demonstrari de entibus causatis, et nullum substantiale potest demonstrari, quare etc. Et hoc etiam dicit Avicenna,[d] et Algazel[e] in logicis suis, sicut alibi patet, et prius patuit. Esse igitur in effectu non est de substantia alicuius, sed est consequens ad quodlibet aggregatum ex materia et forma ratione forme, sicut propria passio sequitur ad proprium subiectum et est eiusdem communitatis cum illo.

Et est intelligendum quod contingit scire quod quid est ⟨dupliciter: potest scire quod quid est⟩ absolute et in se non intelligendo si est, sicut contingit scire prius non sciendo posterius; aut potest scire quod quid est alicuius sub relatione ad esse in effectu cuius causa est, et tunc intelligendo ⟨quod⟩ quid est oportet cognoscere causatum ab eo, quia non contingit cognoscere causam sub ratione cause nisi intelligatur effectus.

Unde per illud apparet ad rationem, quia cum arguitur quod quid est se habet ad si est, sicut propter quid ad quia est, sicut dicit Aristoteles[f] in secundo libro, dico quod propter quid est prius quia est, sicut causa est prior suo effectu, et contingit intelligere prius non intelligendo posterius,

---

[a] Aristoteles, *AnPo*, II, 8, 93 a 16–22   [b] Aristoteles, *AnPo*, II, 7, 92 b 12–18   [c] Aristoteles, *AnPo*, II, 7, 92 b 12–18   [d] Avicenna, *Sufficientia*, I, 3 (1508, f. 3va); *De philosophia prima et scientia divina*, VIII, 3 (ed. Van Riet, pp. 395–97)   [e] Al Ghazali, *Logica* (ed. Lohr, p. 247)   [f] Aristoteles, *AnPo*, II, 1, 89 b 5–35

---

8 si] quid M   18 etiam] cum M   24 se] re M

1.2 Furthermore, just as because-of-what is related to that-it-is-so, so what-it-is is related to if-it-is, according to Aristotle in *Posterior Analytics* II. But if it is not possible to know that-it-is-so, then it is not possible to know because-of-what; therefore etc.

2 On the other hand, Aristotle says that one can know about a triangle what-it-is and still can doubt if-it-is, therefore etc.

3 In response to this it should be understood that the question of if-it-is can ask two things, for it can ask if-something-is absolutely or if-it-is in actual effect. If the question is put in the first way, then the question of what-it-is presupposes the question of if-it-is, since what is primarily posited in the essence of something is the nature of being, and therefore the account of being appears primarily in the concept of any thing. But if the question of if-it-is is put in the second way, then the question of what-it-is does *not* presuppose the question of if-it-is, since being in actual effect is not in the account of entities that are caused, according to Aristotle here, who says next that the question of if something is, or its being, can be demonstrated of entities that are caused, but that nothing substantial can be demonstrated; therefore etc. Avicenna and Alghazali also say this in their logics, as is clear elsewhere and was made clear earlier ⟨q. 49⟩. Therefore being in actual effect does not belong to the substance of a thing. Instead it follows from any aggregation of matter and form, by reason of its form, just as a proper attribute follows from its proper subject and belongs to the same generality as its subject.

It should also be understood that it is possible to know an essence in two ways, without qualification and in itself, without understanding if-it-is, just as it is possible to know something prior without knowing something posterior. Or one can know the essence of a thing under its relation to the being in actual effect of which it is the cause, and then in understanding what-it-is one must cognize what it causes, since one does not cognize a cause, considered as cause, unless one understands its effect.

Hence it is clear what to say to the argument, for when it is argued that what-it-is is related to if-it-is as because-of-what is related to that-it-is-so, as Aristotle says in Book II, I reply that because-of-what-something-is is prior to that-it-is-so, just as a cause is prior to its effect, and it is possible to

ideo contingit intelligere et scire propter quid est sine relatione ad quia ⟨est⟩; similiter dico de quid est respectu si est; quare etc.

Et patet ad rationes, procedunt enim viis suis. Unde Aristoteles in hoc libro quandoque dicit quod non contingit intelligere quid est nisi intelligatur si est, et quandoque dicit oppositum, intelligendo hoc diversimode, sicut expositum est.

Ad 1.1 Ad rationem primam, est dicendum: quod non est simpliciter non contingit scire, ens tamen apud animam, tale contingit scire, ut figmenta. De talibus tamen non contingit scire quid est simpliciter, et ita de his contingit scire quid per nomen significatur.

Ad 1.2 Ad aliud, [dicendum] quod non contingit scire propter quid nisi sciatur quia, dicendum ad hoc per interemptionem: quamvis non contingat scire propter quid sub ratione cause nisi sciatur quia, tamen contingit scire ipsum propter quid secundum se nihil sciendo de quia in actu. Similiter dico de quod quid est respectu eius si est, et hoc apparet, quia quorumcumque est definitio eorum est quod quid est, quod sciri potest. Sed universalium est definitio. Et cum possibile sit aliquid esse universale absque eo quod sit in re, ut domus heptangula non existit in re, existit tamen in suis partibus, quod quamvis composite non sint, compossibiles tamen sunt, iterum, etiam nec ei prohibitum est esse, et tale esse sufficit ad rationem universalis, unde ad cognoscendum quid est non necesse est scire si est presentialiter, quia si sic non contingeret aliquid scire de impressionibus, et sic periret scientia meteororum. Sed cognoscentem quid est necesse est cognoscere si est absolute, id est necesse est scire si res nature fuit per sua principia et causas tantum, etiam ei non est prohibitum esse, et sic scitur eclipsis ipsa non existente; quare etc.

---

33 sine] in M   34 si] quia M   45 sciatur] tamen *add. sed canc.* M   47 de] quod M ‖ si] quod quid M   48 quod²] quia M

understand the prior while not understanding the posterior. Therefore it is possible to understand and know because-of-what, without any relation to the fact that-it-is-so. I reply in the same way concerning what-it-is with respect to if-it-is; therefore etc.

And it is clear how to reply to the arguments, since they go through in their own way. And thus Aristotle in this book sometimes says that understanding what something is occurs only if one understands if-it-is, whereas sometimes he says the opposite, because he understands these claims in different ways, as we have explained.

Ad 1.1 In reply to the first argument, it should be understood that it is not possible to know without qualification something that does not exist. But it is possible to know something with being in the soul, that is, a figment. It is not, however, possible to know the essences of such beings without qualification, and so what is possible to know about these beings is what their name signifies.

Ad 1.2 As to the second—that it is not possible to know because-of-what-it-is unless it is known that-it-is-so—one should reply by rejecting the claim: although it is not possible to know because-of-what-something-is under the aspect (*ratio*) of a cause, unless one knows that-it-is-so, nevertheless it is possible to know because-of-what-something-is, knowing that in its own right, while knowing nothing about whether-it-is-so in actual effect. I claim likewise for an essence with respect to if-it-is. This is clear, since of whatever there is a definition there is an essence that is able to be known. But definitions are of universals, and it is possible for something to be a universal without existing in reality. (For instance, a seven-sided house does not exist in reality, but it exists in its parts which, although they are not put together, are nevertheless able to be put together and, in addition, nothing prevents it from existing. Being such as this suffices for the account of a universal.) Therefore, in order to cognize an essence, it is not necessary to know whether it exists at the moment, for if this were necessary then it would not be possible to know something from its impressions, and so the knowledge of astronomical phenomena would perish. But one who cognizes what-it-is does necessarily cognize if-it-is without qualification. That is, it is necessary to know whether a thing in nature has existed through its principles and causes alone, and also that it is not prevented from existing. And thus an eclipse is known even when it does not exist. Therefore etc.

⟨Questio 59⟩

Queritur utrum contingat scire quid est ycocervus.

1.1 Et arguitur quod sic, quia si non, hoc est quia non est ens penitus, et hoc non est verum, quia de eo quod non est non contingit aliquid enuntiare, per Avicennam[a] in *Metaphysica* sua; sed de chimera sive ycocervo aliquid enuntiat intellectus. Aut quia est ens occultum, et hoc non est verum, per Commentatorem[b] II *Metaphysice*, qui dicit quod nullum ens est occultum intellectui. Quare etc.

1.2 Item, quod non est non movet intellectum; sed chimera movet intellectum; quare etc.

2 Oppositum vult Aristoteles[c] dicens quod non contingit aliquid intelligere de ycocervo; quare etc.

3 Dicendum, sicut dicit Avicenna: de eo quod non est non contingit aliquid intelligere vel enuntiare vel significare; si ergo ycocervus habeat aliquod esse cognitum vel significatum, habet aliquam rationem entis quamquam debilem. Unde privationes, secundum Aristotelem[d] in principio IV *Metaphysice*, aliquam rationem entis habent saltem per attributionem ad suos habitus. Sed esse chimere est esse suarum partium extra animam, tamen nomen chimere illas partes non significat sed compositionem partium, quam impossibile est esse unius in rerum natura, ideo dicit Aristoteles[e] quod hoc non contingit intelligere. Si tamen asinus non esse⟨t⟩, adhuc de eo contingeret intelligere quid est, quia non est sibi prohibitum esse. De chimera igitur non contingit intelligere quid est, sed quid est quod significatur per nomen contingit intelligere de chimera. Et sic contingit intelligere de negatione et privatione et infinito.

Ad 1.1 Ad argumentum primum, dicendum quod [non est] non contingit intelligere quid est de ycocervo, | quia sibi prohibitum est esse, de eo tamen contingit intelligere quid est quod significatur per nomen, et sic posset definitionem habere.

M 112ra

---

[a] Avicenna latinus, *De philosophia prima et scientia divina*, I, 5 (ed. Van Riet, pp. 36–37) [b] Averroes, *Commentarium magnum in Metaphysicam Aristotelis*, II, 1, comm. 1 (1562, ff. 28M–29A) [c] Aristoteles, *AnPo*, II, 7, 92 b 4–11 [d] Aristoteles, *Met.*, IV, 2, 1004 a 14–16 [e] Aristoteles, *AnPo*, II, 7, 92 b 4–8

12–13 aliquid] scire *add. sed exp.* M   19 ideo] idem M   20 dicit] *iter.* M

Question 59.

It is asked whether it is possible to know what a goat-stag is.

1.1 It is argued that it is, since if not, this is because it is not a being at all, and this is not true since it is not possible to assert anything concerning what does not exist, according to Avicenna in his *Metaphysics*, and yet the intellect does assert something about a chimera or goat-stag. Or it is because it is an occult entity, but this is not true according to the Commentator, on *Metaphysics* II, who says that no entity is occult for the intellect. Therefore etc.

1.2 Furthermore, what does not exist does not move the intellect, but a chimera moves the intellect, therefore etc.

2 On the other hand, Aristotle says that it is not possible to understand anything about a goat-stag, therefore etc.

3 It should be replied as Avicenna does: that it is not possible to understand anything about what does not exist, or to assert or signify anything about it. If, then, a goat-stag has some cognized or signified being then it has some account of being, even if it is a diminished one. Hence privations, according to Aristotle at the start of *Metaphysics* IV, have some account of being, at least through an attribution to the states of which they are privations. But the being of a chimera is the being of its parts outside the soul. The name 'chimera', however, does not signify those parts but their composition, which is impossible to belong to some one thing in the world. Thus Aristotle says it cannot be understood. But if a donkey were not a being, it would still be possible to understand what-it-is, since being is not prevented from it. Therefore it is not possible to understand of a chimera what-it-is, but it is possible to understand of a chimera what it is that is signified by its name. And the same is possible to understand about negation, privation, and the infinite.

Ad 1.1 In response to the first argument, it should be replied that it is not possible to understand what-it-is about a goat-stag, because being is prohibited from it. But it is possible to understand of it what it is that is signified by its name, and so it can have a definition.

Ad 1.2 Ad aliud, dicendum quod illud quod nullo modo est non movet intellectum; sed chimera, quamvis sit non ens in rerum natura, est tamen ens apud animam; quare etc.

⟨Questio 60⟩
Queritur utrum definitio significet esse illius cuius est definitio.

1.1 Et videtur quod sic, quia quod significat posterius significat prius; sed quod quid est posterius est esse, quia questio quid est posterior est questione si est, ergo, cum definitio significet quod quid est, significat esse; quare etc.

1.2 Item, quandocumque aliqua duo una cognitione accipiuntur, quidquid significat unum et aliud; sed quid est et quia est eadem cognitione cognoscuntur, per Aristotelem[a] VI *Metaphysice*; quare etc.

2 Oppositum dicit Philosophus in littera, dicens quod definitio non significat quia est nec quia possibile est esse, quia si definitio significaret esse, tunc scientem definitionem alicuius non contingeret querere si est illud definitum; nunc autem dicit Philosophus:[b] contingit querere si est circulus.

3 Dicendum quod definitio potest sumi in comparatione ad definitum de quo enuntiatur, et sic significat definitum esse, et secundum hoc dicit Aristoteles[c] in II huius quod definitio enuntiata de aliquo definito dicit esse affirmative et universaliter. Aut potest definitio accipi secundum se, et sic nec significat esse nec non esse in effectu; secundum hunc modum intelligit Aristoteles[d] in primo huius quod definitio non est suppositio indicans esse vel non esse, quia definitio non significat esse vel non esse.

Item, definitio solum indicat quod quid est rei; sed esse in effectu non est de ipso quod quid est, sed consequens ad ipsum; ideo etc.

---

[a] Aristoteles, *Met.*, VI, 1, 1025 b 15–17   [b] Aristoteles, *AnPo*, II, 7, 92 b 19–25   [c] Aristoteles, *AnPo*, II, 3, 90 b 4–5   [d] Aristoteles, *AnPo*, I, 2, 72 a 19–24

---

19 quod] cum M   20 quia] sed M

Ad 1.2 In response to the second argument, it should be replied that that which in no way exists does not move the intellect. But a chimera, although it is not an entity in the natural world, is still an entity within the soul; therefore etc.

Question 60.

It is asked whether a definition signifies the being of that of which it is a definition.

1.1 It seems that it does, for what signifies the posterior signifies the prior, but essence is posterior to being, since the question of what-it-is is posterior to the question of if-it-is; therefore, since the definition signifies the essence, it signifies being; therefore etc.

1.2 Furthermore, whenever two things are received in one cognition, whatever signifies the one also signifies the other. But what-it-is and that-it-is-so are cognized by the same cognition, according to Aristotle in *Metaphysics* VI; therefore etc.

2 The Philosopher says the opposite in the text at hand, claiming that a definition does not signify that-it-is-so nor that it is possible for it to be so. For if a definition were to signify being then one who knows the definition of something could not ask whether the thing defined exists. But the Philosopher says that one can ask whether there is a circle.

3 It should be replied that a definition can be taken in relation to the thing defined of which it is asserted, and in this way it signifies that the thing defined exists. And in this way Aristotle says in the second book of the *Posterior Analytics* that a definition asserted about some defined thing indicates being affirmatively and universally. Or a definition can be taken with respect to itself, and in this way it signifies neither being nor non-being in actual effect. Aristotle understands it in this way in the first book: that a definition is not a supposition indicating being or non-being, because a definition does not signify being or non-being.

Further, a definition indicates only the essence of a thing, but being in actual effect does not belong to a thing's essence but instead comes after it; therefore etc.

Item, potentia et actus sunt eiusdem generis; si igitur definitio non significat actum essendi definiti, ergo nec eius potentia, et secundum hoc dicit Aristoteles quod definitio non significat esse nec posse esse definiti; quare etc.

Ad 1.1 Ad primum argumentum, dicendum quod esse in effectu non est prius eo quod quid est, quamvis esse simpliciter prius sit, sicut prius patuit; quare etc.

Ad 1.2 Ad aliud, dicendum quod quod quid est et si est non eadem acceptione in cognitionem accipiuntur; quod quid est enim est causa huius quod est si est, et non eadem cognitione cognoscitur causa et effectus. Vel potest dici quod dictum Aristotelis est intelligendum de quia est simpliciter, quando dicit quod quid est et quia est eadem cognitione cognoscuntur, non autem de esse in effectu.

⟨Questio 61⟩

Queritur utrum significatum dictionis possit demonstrari sive ostendi.

1.1 Et videtur quod sic, quia eorum quorum est aliqua causa est demonstratio; sed quod hoc nomen significet habet aliquam causam, cum hoc non sit ens per se; ergo poterit demonstrari per suam causam.

1.2 Item, mediatorum est demonstratio; sed hoc nomen significare hoc est mediatum, quia predicatum non cadit in definitione subiecti; quare etc.

2.1 In oppositum est Aristoteles.[a]

2.2 Et arguitur ratione, quia illud quod supponitur in scientia tamquam primum non demonstratur in illa; sed quid est illud quod per nomen significatur supponitur et est tamquam primum in omni scientia; ergo etc. Et hoc est considerandum propter rationem iam tactam, quia illud quod supponitur non ostenditur; sed significationes vocabulorum supponuntur in omni scientia; igitur etc. Unde si quis neget principia alicuius scientie, semper currendum est ad significationes vocabulorum et per eas

[a] Aristoteles, *AnPo*, II, 7, 92 a 34–b 38

5 nomen] enim M   10 supponitur] *iter.* M

Further, potentiality and actuality belong to the same genus. So if a definition does not signify the act of being of the thing defined, then neither does it signify its potential ⟨for being⟩. And in this way Aristotle says that a definition signifies neither the being nor the possibility of being of the thing defined; therefore etc.

Ad 1.1 In reply to the first argument, it should be said that being in actual effect is not prior to essence, although being without qualification is prior, as was clear earlier; therefore etc.

Ad 1.2 In reply to the second argument, it should be held that essence and if-it-is are not received into cognition by the same reception, for the essence is the cause of that which it is, if it is, and cause and effect are not cognized by the same cognition. Or, it can be held that Aristotle's dictum is to be understood of whether-it-is-so without qualification—when he says that the essence and that-something-is-so are cognized by the same cognition—but is not to be understood of being in actual effect.

Question 61.

It is asked whether what is signified by an expression (*dictionis*) can be demonstrated or shown.

1.1 It seems that it can, for there is demonstration of things of which there is a cause, but what a certain name signifies has a cause, since this is not a *per se* being; therefore it will be demonstrable through its cause.

1.2 Furthermore, there is a demonstration of whatever has a middle term, but that this name signifies this has a middle term, since the predicate does not fall under the definition of the subject; therefore etc.

2.1 On the other side is Aristotle.

2.2 That side is also argued for by reason, since that which is presupposed in a science as first is not demonstrated within it. But what it is that is signified through the name is presupposed as first in every science. Therefore etc. And this should be considered in light of an argument already touched upon, that that which is presupposed is not shown, but the significations of words are presupposed in every science; therefore etc. Accordingly, if anyone denies the principles of a science, the discussion

ostendit artifex prima principia; et si quis neget significationes vocabulorum, cum tali non est ulterius disputandum magis quam cum planta, per Aristotelem[a] IV *Metaphysice*; quare etc.

3 Dicendum igitur quod illud quod debet demonstrari oportet habere causam nobis notam et non indeterminatam sed certam. Sed quod hoc nomen significet hoc, huius est aliqua causa, sed illa est indeterminata, scilicet voluntas imponentis, usus auctorum, et quia voluntas ex infinitis posset moveri et a tali impositione impediri, cum ipsa sit ad placitum, ideo non habemus causam determinatam. Ideo nulla scientia potest demonstrare significationes vocabulorum, et hoc demonstratione a priori, ostendi tamen possunt a posteriori ut per usus auctorum. Sed apparet quod non cuiuslibet habentis aliquam causam est demonstratio, sed habentis causam certam et determinatam, similiter non omnium mediatorum est demonstratio, sed certam causam habentium.

Ad 1.1 et 1.2 Et per hoc patet ad rationes.

⟨Questio 62⟩

Queritur circa illud capitulum: *Iterum autem speculandum* (93 a 1), et primo utrum aliqua causa sit idem cum eo cuius est causa.

1.1 Et videtur quod sic, quia illud quod ponitur in definitione alicuius est idem illi; sed alique cause ponuntur in definitione eorum quorum sunt cause, unde Aristoteles[b] in IV *Metaphysice* commendat Archistam, quia accipit omnes causas in definitione sua; ergo etc.

1.2 Item, hoc idem arguitur de causis, particulariter de causis materialibus et formalibus que videntur esse idem cum eo cuius sunt cause, quia ingrediuntur definitiones et substantiam rei. Et hoc idem dicit Aristotelem[c] hic; dicit | enim quasdam esse causas easdem cum re et quasdam non. Cause enim intrinsece sunt idem, sed cause extrinsece non sunt idem; quare etc.

M 112rb

---

[a] Aristoteles, *Met.*, IV, 4, 1006 a 15; 1008 b 11–12   [b] Aristoteles, *Met.*, VIII, 2, 1043 a 21–25
[c] Aristoteles, *AnPo*, II, 9, 93 b 21–22

---

18 demonstrari] demonstrat M   25 tamen] non M   26 aliquam] alteram M

must always run to the significations of the words. Also, it is through these that the artisan shows the first principles. Also, if anyone denies the significations of the words, one should not dispute any further with him, any more than with a plant, as Aristotle puts it in *Metaphysics* IV. Therefore etc.

3 It should therefore be replied that that which we are to demonstrate must have a cause known to us that is not indeterminate but instead certain. Now there is some cause for this name's signifying this, but it is an indeterminate cause, namely the will of the one who imposes the meaning or the custom of authors. And since the will can be moved by infinitely many things and can be impeded from such an imposition, since it does as it pleases, we therefore do not have a determinate cause. Therefore no science can demonstrate the significations of words through an *a priori* demonstration. They can, however, be shown *a posteriori*, for instance through the custom of authors. But it is clear that there is not a demonstration for *anything* that has a cause, but only for what has a certain and determinate cause. Similarly, there is not a demonstration of everything with a middle term, but only of those that have a certain cause.

Ad 1.1 and 1.2 And from this it is obvious how to reply to the arguments.

Question 62.

With respect to the chapter that begins 'However, it must be examined again' (93 a 1), it is first asked whether there is any cause that is the same as that of which it is the cause.

1.1 It seems that there is, for that which is posited in the definition of something is the same as it, but some causes are posited in the definition of those things they are causes of. Hence Aristotle in *Metaphysics* IV commends Archytas because he accepts all causes in his definition. Therefore etc.

1.2 Furthermore, an argument for the same conclusion comes from causes, particularly material and formal causes, which seem to be the same as that of which they are causes, since they enter into definitions and the substance of a thing. And Aristotle says this here, for he says that some causes are the same as a thing and some are not, since intrinsic causes are the same whereas extrinsic causes are not. Therefore etc.

2 Oppositum arguitur: illud non est idem alicui quod ipsum subsequitur; sed causa precedit illud cuius est causa, ut patet per definitionem cause, que est ad cuius esse sequitur aliud, ut causatum; ergo causa et causatum non sunt idem.

3 Dicendum quod nulla causa simpliciter est idem cum eo cuius est causa, quia causa precedit illud cuius est causa; sed nihil precedit se ipsum; ergo etc.

Sed est intelligendum quod 'idem' dicitur dupliciter, scilicet subiecto et ratione. Et similiter 'diversum' dicitur dupliciter, scilicet subiecto et ratione [diversa], diversa subiecto ut Sortes et Plato, ⟨diversa ratione sed⟩ eadem subiecto ut albedo et dulcedo in lacte. Unde in aliquibus causa et effectus sunt idem subiecto, ut materia et forma sunt idem subiecto cum eis quorum sunt cause ⟨sed sunt diversa ratione⟩. Unde quamvis homo componatur ex materia et forma, alia tamen est ratio hominis et alia ratio materie et alia ratio forme, et hoc apparet si consideremus ad propria eorum. Si autem consideremus ad causam agentem et effectum eius, tunc aliquando differunt subiecto et conveniunt ratione, ut quando eadem est forma generantis et generati, ut cum homo generatur ex homine et calidum ex calido; et aliquando causa agens et effectus non conveniunt in forma, conveniunt tamen in subiecto. De causa autem finali ultima, quam voco operationem rei, manifestum quod ipsa differt a re cuius est causa in ratione; causa autem finalis proxima, quam voco formam rei, se habet ad rem sicut forma se habet ad illud cuius est forma. Unde quedam cause per Aristotelem sunt idem rei, ut forma et materia, que per causam intrinsecam non demonstrantur, unde dicit Aristoteles[a] quod definitio data per ultimam formam demonstrari non potest, causa tamen materialis forte demonstrari potest per formam, sed cause extrinsece non sunt idem cum eo cuius sunt cause, ut agens et finis.

Ad 1.1 Ad primum argumentum, dicendum quod non omne quod ponitur in definitione alicuius est idem sibi in essentia, quia definitiones quedam dantur per additamenta, ut definitiones accidentium, et quedam per accidentia, sicut quorumdam accidentium, et in his non est verum quod assumptum est, sed solum in definitione substantie.

[a] Aristoteles, *AnPo*, II, 9, 93 b 21–22

41 argumentum] arguitur M   45 substantie] natura M

2 On the other hand, it is argued that a thing is not the same as anything that follows upon it, but a cause precedes that of which it is the cause, as is clear from the definition of cause, which is that upon the being of which another follows as that which is caused. Therefore cause and caused are not the same.

3 It should be replied that no cause is without qualification the same as that of which it is the cause, for the cause precedes that of which it is the cause, but nothing precedes itself; therefore etc.

But it should be understood that 'the same' is said in two ways, namely in subject and in account (*ratione*). In the same way, 'different' is said in two ways, in subject and in account. Things such as Socrates and Plato are different in subject, and things such as whiteness and sweetness in milk are the same in subject but different in account. Hence in some cases the cause and the effect are the same in subject, as matter and form are the same in subject with those things of which they are the causes. Hence, although a human being is composed of matter and form, still, the account of human being is distinct from the account of the matter and the account of the form. And this is apparent if we consider their properties. But if we consider the agent cause and its effect, then sometimes they differ in subject and agree in account, as when the form of the generator and the generated are the same—for instance, when a human being comes to be from a human being and heat from heat. Other times the agent cause and the effect do not agree in form but agree in subject. But concerning the ultimate final cause, which I call the operation of the thing, it is obvious that it differs in account from the thing of which it is the cause. But the proximate final cause, which I call the form of a thing, is related to the thing as form is related to that of which it is the form. Hence certain causes are, according to Aristotle, the same as the thing, such as form and matter, which are not demonstrated through an intrinsic cause. Hence Aristotle says that a definition given through the ultimate form cannot be demonstrated, but the material cause can perhaps be demonstrated through the form. But the extrinsic causes, the agent and the end, are not the same as that of which they are causes.

Ad 1.1 In reply to the first argument, it should be said that not everything that is posited in a definition of a thing is the same as it in essence, since some definitions are given through additions, such as the definitions of accidents, and some definitions are given through accidents, as of certain accidents. In these cases what the argument assumes is not true. It is true only for the definition of a substance.

Ad 1.2 Ad aliud, [sicut] prius patuit in dicendo que sunt cause eedem rebus quarum sunt cause et que non, et hoc dico secundum rem; planum enim est quod cause sub ratione cause differunt ab effectibus saltem secundum rationem.

⟨Questio 63⟩

Queritur utrum aliqua causa possit demonstrari, sicut dicit Aristoteles.[a]

1.1 Et videtur quod non, quia immediatorum non est demonstratio; sed causa per se est immediata illius cuius est causa; ergo etc.

1.2 Item, si causarum per se sit demonstratio, ergo cuiuslibet cause erit demonstratio; et ita demonstrationes procedent in infinitum, cum omnis demonstratio fiat per causam, quod est inconveniens; quare etc.

2 Oppositum dicit Aristoteles,[b] dicens causas mediatas demonstrabiles esse.

3 Advertendum quod dupliciter est demonstratio, scilicet propter quid, et talis procedit per causam proximam, et demonstratio quia, et talis procedit a causa remota ad propinquam vel ab effectibus ad causam. Unde in aliquibus effectus sunt notiores suis causis, et tales cause possunt demonstrari per effectus. Similiter et demonstratione propter quid possunt quedam cause demonstrari, quia cause sibi invicem sunt cause, per Aristotelem[c] II *Physicorum*, unde cum causa formalis sit prior secundum rationem quam causa materialis, quia materia fit propter formam, et ideo causa materialis et eius definitio demonstrari potest per causam formalem et eius definitionem. Cum finis sit causa causarum, ideo per causam finalem possunt omnes alie cause demonstrari.

Ad argumenta in oppositum.

Ad 1.1 Ad primum, dicendum quod non omnes cause immediate insunt rei nisi in genere subiecti, quedam tamen sunt cause esse rei per alias, et ideo quedam per alias demonstrari possunt.

---

[a] Aristoteles, *AnPo*, II, 9, 93 b 26–28 [b] Aristoteles, *AnPo*, II, 9, 93 b 26–28 [c] Aristoteles, *Phys.*, II, 3, 195 a 8–9

---

9 advertendum] advertandum M   14 cause¹] causa M ‖ cause²] causa M   17 causa] formalis *add. sed* malis *exp.* M   ‖ eius] cuius M   22 cause] causa M

Ad 1.2 The reply to the other argument was clear earlier, in saying which causes are the same as that of which they are causes, and which are not. And I say this as they are in reality, for it is obvious that causes, when characterized (*sub ratione*) as a cause, differ from their effects, at least according to their account.

Question 63.

It is asked whether a cause can be demonstrated, as Aristotle says.

1.1 It seems not, since there is no demonstration of immediate things, but a *per se* cause is the immediate cause of that of which it is the cause; therefore etc.

1.2 Furthermore, if there is a demonstration of *per se* causes, then there will be a demonstration of every cause, and so demonstrations will proceed to infinity, since every demonstration is made through a cause. This is absurd, therefore etc.

2 On the other hand, Aristotle says that middle causes are demonstrable.

3 It should be noted that there are two kinds of demonstration: demonstration why (*propter quid*), which proceeds from a proximate cause, and demonstration that (*quia*), which proceeds from a remote cause to one that is nearer, or from effects to a cause. Hence, in some cases the effects are better known than their causes, and such causes can be demonstrated through their effects. Similarly, some causes can be demonstrated by a demonstration why, since the causes are causes of each other, according to Aristotle in *Physics* II. Hence, since the formal cause is prior in account to the material cause, for matter comes about because of form, so the material cause and its definition can be demonstrated through the formal cause and its definition. And since the end is the cause of causes, all other causes can be demonstrated through the final cause.

In response to the arguments.

Ad 1.1 To the first it should be replied that not every immediate cause inheres in a thing, unless the cause happens to be in the genus of the subject, but there are some causes of the being of a thing through others, and therefore some can be demonstrated through others.

Ad 1.2 Ad secundum, dicendum quod cause unde causa est non est demonstratio, sed cause unde mediata est per causam priorem est demonstratio, et ideo non concludit dicta ratio.

⟨Questio 64⟩

Queritur utrum contingat scire per causam.

1 Et videtur quod non, quia sic nihil sciretur nisi sciretur causa prima; sed scire causam primam perfecte est impossibile; quare etc.

2.1 Oppositum dicit Aristoteles.ᵃ

2.2 Et apparet per definitionem eius quod est scire, que est per causam cognoscere,ᵇ etc.

3 Ad questionem dico quod contingit scire per causam, quia sicut se habet aliquid ad esse similiter ad cognitionem; sed res habet esse per suam causam; ergo per eam cognoscitur.

Sed est intelligendum quod quamvis eadem sint principia essendi et cognoscendi, non tamen sub eadem ratione, quia principia cognitionis sunt universalia et principia essendi sunt particularia. Et sicut omnia habent principium essendi a primo principio, similiter et cognoscendi, sed differt aliqualiter, quia causa | prima, inquantum est principium cognoscendi omnium, accipi debet in universali sub ratione entis, et ita patet quod causa prima inquantum particulariter accepta est non est a nobis cognoscibilis, inquantum tamen in universali accipitur maxime cognoscibilis est et omnia in virtute eius cognoscuntur.

Sed tunc quereret aliquis qualiter causa prima potest particulariter accipi secundum quod est principium essendi, cum tamen accipiatur in universali secundum quod est principium cognoscendi, cum ipsa causa essendi omnium sit sicut et cognoscendi. Dicendum quod causa prima universalis est causalitate, sub esse tamen particulari, sed non est universalis predicatione nisi prout sumitur sub ratione entis.

Ad 1 ⟨...⟩

ᵃ Aristoteles, *AnPo*, II, 11, 94 a 20–21   ᵇ Cf. Aristoteles, *AnPo*, I, 2, 72 b 10–13

---

11 ratione] nota de ista propositione quod eadem sunt principia essendi et cognoscendi *add.* M

Ad 1.2 To the second it should be said that of causes as such there is no demonstration, but of causes as mediated through a prior cause there is demonstration, and therefore the stated argument does not succeed.

Question 64.

It is asked whether one can know through a cause.

1 It seems one cannot, because if so then nothing would be known unless it were known through the first cause, but it is impossible to know the first cause perfectly; therefore etc.

2.1 Aristotle takes the opposite view.

2.2 This is apparent through his definition of what it is to know, which is to cognize through a cause, etc.

3 I reply to the question that one can know through a cause, since just as a thing is related to being, so is it related to cognition. But a thing has being through its cause, and therefore it is cognized through its cause.

It should be understood, however, that although the principles of being and cognizing are the same, they are not under the same account, because the principles of cognition are universals and the principles of being are particulars. And just as everything has a principle of being from the first principle, so too it has a principle of cognizing. But they differ in a way, since the first cause, insofar as it is the principle of cognizing for all things, ought to be taken universally under the account of being. And so it is obvious that the first cause, insofar as it is taken as particular, is not cognizable by us, but insofar as it is taken universally, it is most cognizable, and everything is cognized in virtue of it.

But then someone will ask how the first cause can be taken as a particular as it is a principle of being, when it is nevertheless taken universally as a principle of cognizing, since it is the cause of the being of all things, just as it is also the cause of the cognition of all things. It should be replied that the first cause is universal by causality, under particular being; but it is universal by predication only as it is taken under the account of being.

Ad 1 ⟨...⟩[18]

---

[18] The manuscripts lack a response to the initial objections.

⟨Questio 65⟩

Queritur utrum ad cognitionem rei completam exigatur cognitio omnium causarum.

1.1 Et videtur quod sic per Aristotelem[a] I *Physicorum*, ubi dicit quod non opinamur scire donec sciamus omnes causas propinquas et remotas et principia et usque ad elementa; et dicit ibidem Commentator[b] quod doctrina non debet cessare de aliquo ente quousque habeatur universaliter cognitio omnium causarum illius et hoc in quolibet genere; quare etc.

1.2 Item, si scire est causam rei cognoscere, ergo in tantum deficit unusquisque a cognitione rei in quantum deficit a cognitione cause; ergo si contingit aliquam rem simpliciter scire, necesse est omnes causas eius cognoscere.

2.1 Oppositum arguitur: demonstratio propter quid facit simpliciter scire, et tamen non accipit omnes causas conclusionis cuius facit scientiam, sed solum causam efficientem; quare etc.

2.2 Item, definitio est principium sufficiens cognoscendi rem, et tamen non accipit omnes causas, sed intrinsecas solum; quare etc.

2.3 Item, I *Ethicorum*,[c] medicus est causa sufficiens sanitatis, et tamen medicus non inducit sanitatem usque in causam primam, sed in causas particulares, quia ad operandum parum prodest primam causam cognoscere.

3 Intelligendum quod sicut non sunt eadem nobis nota et nota nature, sic nec sunt eedem cause cognitionis nostre et nature rei. In natura enim unumquodque est notum ex illis ex quibus entitatem habet, et quia habet entitatem ex omnibus suis causis, ideo etc. Et quia aliquando effectus sunt nobis notiores suis causis, sicut ista inferiora sunt notiora superioribus, quia magis sensibilia et magis propinqua, ideo nos cognoscimus aliquando causas per effectus, et non semper sunt eadem principia rei et doctrine nostre.

---

[a] Aristoteles, *Phys.*, I, 1, 184 a 10–15 [b] Averroes, *Commentarium magnum in Physicam Aristotelis*, I, 1, comm. 1 (1562, f. 6D) [c] Aristoteles, *NE*, I, 6, 1097 a 7–14

---

27 et²] nature *add. sed del.* M

Question 65.

It is asked whether the complete cognition of a thing requires cognizing all of a thing's causes.

1.1 It seems that it is, for Aristotle in *Physics* I says that we do not consider ourselves to know until we know all the causes, both proximate and remote, and the principles right down to the elements. And the Commentator says in that place that teaching should not stop regarding some entity until a cognition of all the causes is had in every case, for each kind of cause; therefore etc.

1.2 Furthermore, if knowing is cognizing the cause of a thing, then each one lacks cognition of a thing to the extent that he lacks cognition of its causes. Therefore, if it is possible for anyone to know something without qualification then it is necessary that he cognize all of its causes.

2.1 The contrary is argued thus: demonstration of why something is the case produces knowing without qualification and yet it does not include all the causes of the conclusion it produces knowledge of, but only the efficient cause; therefore etc.

2.2 Furthermore, a definition is a sufficient principle for cognizing a thing, and yet it does not include every cause, but only the thing's intrinsic causes; therefore etc.

2.3 Furthermore, according to *Ethics* I, a doctor is a sufficient cause of health, and yet a doctor does not produce health through a cognition that goes all the way to the first cause, but only to the particular causes, since cognizing the first cause provides little help in performing his function.

3 Reply. It should be understood that just as what is known to us is not the same as what is known by nature, so neither are the causes of our cognition the same as the causes of a thing's nature. For in nature each one is known from that from which it has being, and since it has being from all its causes, therefore etc. And since sometimes the effects are better known to us than their causes, just as lower things are better known than higher things, since they are more accessible to the senses and nearer, so we sometimes cognize causes through effects, and the principles of a thing are not always the same as the principles of our teaching.

Ad 1.1 et 1.2 Et per hoc patet ad rationes. Prime enim ostendunt quod ad cognitionem simpliciter habendam exigitur cognitio omnium causarum; sed talis cognitio non est possibilis humano intellectui. Secundum tamen quod dicit Commentator, per demonstrationem habemus nos cognitionem alicuius secundum quod eius cognitio possibilis est intellectui humano, et non secundum naturam rei et non simpliciter. Similiter etiam definitio facit nos scire rem quantum ad eius principia intrinseca, sed talis scientia non est completa rei cognitio, et ideo demonstratio et definitio faciunt scire in aliquo genere cause simpliciter, non tamen absolute simpliciter.

⟨Questio 66⟩

Queritur utrum ad hoc quod aliquid perfecte et simpliciter sciatur oporteat scire omnes causas eius.

1.1 Et videtur quod non, quia demonstratio facit scire simpliciter; sed demonstratio non facit scire per omnes causas coniunctim, sed divisim per Aristotelem[a] hic; ideo etc.

1.2 Item, si demonstratio faceret scire per quamlibet causam, tunc nihil possit demonstrari nisi cognosceretur causa prima, quod est inconveniens; ideo etc.

2 Oppositum dicit Aristoteles[b] et Commentator[c] I *Physicorum*.

3 Ad hoc dicunt aliqui quod ad hoc quod aliquid sciatur perfecte oportet cognoscere omnes causas in genere, non tamen oportet cognoscere omnes causas eius extrinsecas sive extra genus, ut cum mathematicus velit aliquid scire in genere quantitatis, oportet cognoscere omnes causas eius in illo genere, et hoc sufficit.

Sed hoc non est verum, quia nihil perfecte cognoscitur quousque cognoscantur omnes cause eius, unde si aliquod ens in genere habeat causas extra genus oportet illas cognoscere, quia sicut unumquodque se habet ad

---

[a] Aristoteles, *AnPo*, II, 11   [b] Aristoteles, *Phys.*, I, 1, 184 a 12–14   [c] Averroes, *Commentarium magnum in Physicam Aristotelis*, I, 1, comm. 1 (1562, f. 6D)

---

35 intrinseca] extrinseca *ante corr.* M   16 aliquod] aliquid M

Ad 1.1 and 1.2 Through this it is clear how to reply to the arguments. The initial arguments show that a cognition of every cause is needed for the cognition without qualification of something. But such a cognition is not possible for the human intellect. According to what the Commentator says, however, we have cognition of something through demonstration in such a way that its cognition is possible for the human intellect, although not according to the thing's nature and not without qualification. And in the same way, a definition makes us know a thing with respect to its extrinsic principles, but such knowledge is not a complete cognition of a thing. And so demonstration and definition produce knowledge with respect to some kind of cause without qualification, but not absolutely without qualification.

Question 66.

It is asked whether it is necessary to know all of the causes of something in order to know it perfectly without qualification.

1.1 It seems that it is not, for demonstration produces knowing without qualification, but demonstration does not produce knowing through all the causes taken together, but separately, according to Aristotle here, therefore etc.

1.2 Furthermore, if demonstration produced knowing through every cause, then nothing could be demonstrated unless the first cause were cognized, which is absurd. Therefore etc.

2 Aristotle maintains the opposite, as does the Commentator at *Physics* I.

3 Some hold in response to this that for something to be known perfectly every cause within its genus must be cognized, but it is not necessary to cognize every one of its extrinsic causes—that is, those that fall outside its genus. So, for instance, when mathematicians wish to know something in the genus of quantity, it is necessary for them to cognize all of its causes in that genus, and this suffices.

But this reply is not true, for nothing is cognized perfectly until all of its causes are cognized. Hence, if some being within a genus has causes outside that genus then they have to be cognized, for just as each thing

esse, ita se habet ad esse verum; sed esse verum est esse cognitum; quare sicut unumquodque se habet ad esse, ita et ad cognitionem rectam. Si igitur aliqua sint cause rei extrinsece in essendo, ergo et in cognoscendo, et sicut contingit devenire ad causam primam simpliciter in essendo, ita et in cognoscendo, et sicut omnia sunt in virtute primi, ita etiam omnia in virtute eius cognoscuntur. Cause tamen in essendo sunt particulares, in cognoscendo universales; unde causa prima est principium | cognitionis humane sub ratione universali, principium tamen primum sub esse particulari est finis cognitionis nostre.

Ad argumenta.

Ad 1.1 Ad primum, dicendum quod demonstratio facit scire simpliciter; sed in demonstratione oportet multa supponere, ut causam materialem, de qua supponitur esse; unde quamvis demonstratio non accipiat omnes causas pro medio, per demonstrationem tamen habemus cognitionem omnium causarum.

Ad 1.2 Ad secundum, ⟨dictum est⟩ prius, scilicet qualiter contingit primam causam cognoscere, quoniam sub esse universali, et qualiter non, quoniam sub esse particulari.

⟨Questio 67⟩

Queritur utrum materia sit cognoscibilis.

1.1 Et videtur quod sic, quia omne illud quod ponitur in definitione alicuius est cognoscibilis; sed materia ponitur in definitione alicuius; ergo etc.

1.2 Item, omne quod habet aliquam rationem entis est scibile secundum quod rationem entis habet; sed materia prima habet aliquam rationem entis; ergo etc.

2 Oppositum dicit Aristoteles[a] super I *Physicorum*, ubi ipse dicit quod materia cognoscibilis est per analogiam ad formam; et Aristoteles[b] V *Metaphysice* dicit quod materia cognoscitur per privationem; quare etc.

3 Dicendum quod prima materia cognoscibilis est, cum per causam materialem aliquando demonstretur effectus; et principium demonstrationis

---

[a] Aristoteles, *Phys.*, I, 7, 191 a 9–12   [b] Aristoteles, *Met.*, VII, 3, 1029 a 10–20

stands to being, so it stands to being true. But being true is being cognized, and therefore just as each thing stands to being, so it stands to correct cognition. If, then, there are some extrinsic causes of a thing in being, then so too in cognizing. And just as it is possible to arrive at the first cause without qualification in being, so also in cognizing. And just as all things exist in virtue of the first cause, so also all things are cognized in virtue of the first cause. But causes in being are particulars and are universals in cognizing. Therefore the first cause is the principle of human cognition under the aspect (*ratio*) universal, but the first principle under particular being is the goal of our cognition.

In response to the arguments.

Ad 1.1 In reply to the first, it should be maintained that demonstration produces knowing without qualification, but in a demonstration one must presuppose many things, for instance, the material cause, the existence of which is presupposed. Hence although a demonstration does not take every cause as its middle term, still we have cognition of every cause through demonstration.

Ad 1.2 In reply to the second it should be said, as before, that the first cause can be cognized in a way, under universal being, and in a way it cannot, under particular being.

Question 67.

It is asked if matter is cognizable.

1.1 It seems that it is, for everything included in a thing's definition is cognizable; but matter is included in a thing's definition; therefore etc.

1.2 Furthermore, everything that has some account of entity is knowable inasmuch as it has that account of being; but prime matter has some account of being; therefore etc.

2 Aristotle holds the opposite view in *Physics* I, where he says that matter is cognizable by analogy to form; and Aristotle says in *Metaphysics* V that matter is cognized through privation; therefore etc.

3 It should be replied that prime matter is cognizable, since sometimes the effect is demonstrated through the material cause; but a principle of

scibile est; ideo etc. Sed est intelligendum secundum Aristotelem[a] IX *Metaphysice* quod nihil cognoscitur per se nisi ens actu, scilicet per formam; materia igitur secundum se non est cognoscibilis, sed per formam.

Ad 1.1 et 1.2 Et per hoc patet ad rationes.

⟨**Questio 68**⟩

Queritur utrum aliquid possit demonstrari per causam materialem.

1.1 Et videtur quod non, quia ad illud per quod debet demonstrari aliquid sequitur illud demonstrandum esse; sed ad materiam nihil sequitur esse sed possibile esse; ergo etc.

1.2 Item, causa debet esse notior suo effectu; sed causa materialis secundum se non est nota; quare etc.

2 Oppositum dicit Aristoteles.[b]

3 Dicendum quod causa materialis dicitur de materia que solum est in potentia, et etiam de partibus que habent rationem materie respectu totius. Secundum Aristotelem II *Physicorum*[c] et V *Metaphysice*[d] materiam primam possumus considerare dupliciter, uno modo secundum se solum, alio modo secundum quod est summe disposita ad formam. Per causam autem materialem simpliciter primam et indispositam nihil contingit demonstrare, sicut probant due prime rationes; per causam tamen materialem secundo modo acceptam, scilicet ut est sub dispositionibus debitis, sic contingit per eam aliquid demonstrare, et sic ad materiam sequitur aliquid esse, et etiam sic materia notior est suo effectu. Verum tamen est quod neutro modo intelligit Philosophus hic, sed per causam materialem intelligit partes que habent rationem materie respectu totius. Et quod Aristoteles sic intelligat patet in littera per exemplum eius; unde partes tales sigillatim accepte sunt notiores toto, et ad illas partes sequitur totum; quare etc.

Et per hoc patet ad rationes.

---

[a] Aristoteles, *Met.*, IX, 8, 1049 b 13–17   [b] Aristoteles, *AnPo*, II, 11, 94 a 20-35   [c] Aristoteles, *Phys.*, II, 3, 195 b 9–10   [d] Aristoteles, *Met.*, V, 2, 1013 b 29

demonstration is knowable; therefore etc. But it should be understood, according to Aristotle in *Metaphysics* IX, that nothing is cognized *per se* except being in act, namely, through form. Matter, then, is not cognizable in itself, but through form.

Ad 1.1 and 1.2 The response to the arguments is obvious from this.

Question 68.

It is asked whether something can be demonstrated through a material cause.

1.1 It seems not, for the thing to be demonstrated follows from that through which it must be demonstrated, but no being follows from matter except possible being; therefore etc.

1.2 Furthermore, the cause must be better known than its effect, but the material cause is not known in itself, therefore etc.

2 Aristotle maintains the opposing view.

3 It should be held that the material cause refers to matter that exists only in potentiality and also to parts that have the character (*ratio*) of matter with respect to the whole. According to Aristotle in *Physics* II and *Metaphysics* V, we can consider the first of these matters in two ways; in one way, in itself alone, and in another way, as it is to the highest degree disposed to form. One can demonstrate nothing through the material cause that is primary without qualification and not disposed to form, as the first two arguments prove. But through the material cause taken in the second way, that is, as it falls under the appropriate disposition, it is possible to demonstrate something through it. And in this way some being does follow from matter, and in this way too matter is more known than its effect. Nevertheless, the Philosopher here has in mind neither of these ways. Instead, through 'material cause' he understands parts that have the character of matter with respect to the whole. And that Aristotle has this in mind here is clear from his example in the text. Hence such parts, taken one by one, are better known than the whole, and the whole follows from those parts; therefore etc.

And through this it is obvious how to answer the arguments.

⟨Questio 69⟩

Queritur utrum aliquid possit demonstrari per causam formalem ipsius.

1 Et videtur quod non, quia illud quod demonstratur et per quod aliquid demonstratur differunt; sed forma non differt ab eo cuius est forma; ergo etc.

2.1 In oppositum est Aristoteles.[a]

2.2 Et arguitur ratione, quia illa que demonstrantur demonstrantur per ea que cadunt in ratione eorum, cum medium debeat esse unigeneum extremis; sed illud quod cadit in ratione alicuius habet rationem forme; ideo etc.

3 Dicendum ad hoc quod in demonstratione simpliciter non contingit demonstrare per causam formalem, et hoc passionis, quia causa formalis et passio cuius est causa formalis sunt idem, et nihil potest per se ipsum simpliciter demonstrari. Verum tamen est quod contingit passionem demonstrari per causam formalem eius subiecti, sed per causam formalem huius passionis non contingit passionem demonstrari.

Sed intelligendum est quod forma aliquando dicitur altera pars formalis hominis, aliquando dicitur forma illud quod ponitur in ratione alicuius, ut animal rationale mortale est forma hominis, et sic contingit per causam formalem ultimo dictam passionem demonstrari, quia talis causa formalis notior est eo cuius est causa.

Ad 1 Et per hoc patet ad rationes.

Expliciunt questiones libri *Posteriorum* secundum magistrum Simonem Anglicum bene dicte per universa studia.

---

[a] Aristoteles, *AnPo*, II, 11, 94 a 20–23; 94 a 36

---

8 extremis] extremum M ‖ alicuius] alterius M   14 eius] huius M ‖ sed] scilicet M

Question 69.

It is asked whether anything can be demonstrated through its formal cause.

1 It seems not, for that which is demonstrated and that through which it is demonstrated are different, but the form does not differ from that of which it is the form; therefore etc.

2.1 On the other side is Aristotle.

2.2 This is also argued for by reason, because things that are demonstrated are demonstrated through what falls within their account, since the middle term must be of the same genus as the extremes. But that which falls within the account of something has the account of the form. Therefore etc.

3 It should be replied to this that in a demonstration without qualification it is not possible to demonstrate through the formal cause. This is so for an attribute, since the formal cause and the attribute of which it is the formal cause are the same, and nothing can be demonstrated without qualification through itself. Now an attribute can be demonstrated through the formal cause of its subject, but through the formal cause of an attribute one cannot demonstrate that attribute.

It should be understood, however, that sometimes 'form' refers to a distinct formal part of a human being, and sometimes it refers to what is included in a thing's account—as, for instance, rational mortal animal is the form of a human being. Hence, through a formal cause taken in the latter way, an attribute can be demonstrated, since that sort of formal cause is better known than that of which it is the cause.

Ad 1 And through this the reply to the arguments is obvious.

Here ends the questions on the *Posterior Analytics* of Master Simon the Englishman, questions that are praised throughout all the schools.

Bibliography

Sources

Albert the Great (Albertus Magnus), *Complectens libros 2 de demonstratione, id est Posteriorum analyticorum*, in *Opera Omnia*, vol. 2, ed. A. Borgnet (Paris: Vives, 1890).

—, *De Causis et Processu Universitatis a Prima Causa*, in *Opera Omnia*, vol. 17.2, ed. W. Fauser (Aschendorff: Aschendorff Verlag, 1993).

—, *De V universalibus*, ed. M. Santos Noya (Münster: Ashendorff, 2004, editio digitalis).

Al Ghazali, 'Logica Algazelis. Introduction and Critical Text', ed. C. H. Lohr, *Traditio*, 21 (1965), 223–90.

Ammonius, *Commentaire sur le Peri Hermeneias D'Aristote. Traduction de Guillame de Moerbeke*, ed. G. Verbeke (Louvain: Publications universitaires de Louvain, 1961).

Aristoteles latinus, *Analytica posteriora. Translatio Jacobi*, Aristoteles Latinus IV.1, ed. L. Minio Paluello and B. Dod (Turnhout: Brepols, 1968).

Averroes, *Commentarium magnum in Metaphysicam Aristotelis*, in *Aristotelis metaphysicorum libri XIIII cum Averrois Cordubensis in eosdem commentariis et epitome* (Venetii: Junctas, 1562).

—, *Commentarium magnum in Physica Aristotelis*, in *Aristotelis De physico auditu libri octo cum Averrois Cordubensis variis in eosdem commentariis* (Venetii: Junctas, 1562).

—, *Commentarium magnum in Aristotelis De anima libros*, ed. F. S. Crawford (Cambridge, Mass.: The Medieval Academy of America, 1953).

Avicenna, *Auicene perhypatetici philosophi : ac medicorum facile primi, Opera in luce redacta: ac nuper quantum ars niti potuit per canonicos emendata. Logyca, Sufficientia, De celo et mundo, De anima, De animalibus, De intelligentijs, Alpharabius de intelligentijs, Philosophia prima* (1508, reprinted in Frankfurt am Main: Minerva, 1961).

—, *Liber de philosophia prima sive scientia divina I-IV*, ed. S. Van Riet (Louvain: Peeters, 1977).

Boethius, *The Theological Tractates and The Consolation of Philosophy*, ed. E. K. Rand and S. J. Tester, Eng. tr. S. J. Tester (Cambridge, Mass., and London: Harvard University Press-Heinemann, 1973).

—, *De Topicis Differentiis* (Paris: Patrologia latina 44, 1860).

—, *Boethius's De Topicis Differentiis*, Eng. tr. E. Stump (Ithaca, NY, and London: Cornell University Press, 1978).

Giles of Rome (Aegidius Romanus), *Egidius super libros Posteriorum Aristotelis* (Venetii: Bonetus locatelus, 1488).

—, *De esse et essentia, de mensura angelorum, de cognitione angelorum* (= *Quaestiones disputatae*) (Venetii: imp. per Simonem de Luere, 1503).

—, *In Aristotelis libros De generatione commentaria* (Venetii: H. Scotum, 1505).

—, *Aegidii Romani Theoremata de esse et essentia. Texte précédé d'une introduction historique et critique*, ed. E. Hocedez (Louvain: Museum Lessianum, 1930).

—, *De medio demonstrationis*, ed. I. Pinborg in 'Diskussion um die Wissenschaftstheorie an der Artistenfakultät', *Miscellanea Mediaevalia*, 10 (1976), 240–68.

Maimonides, *Dux Perplexorum* (Paris: Augustinus Justinianus, 1520).

Pophyry (Porphyrium), *Introductio in Aristotelis Categorias a Boethio translatas*, ed. A. Busse (Berlin: De Gruyter, 1962).

Proclus, 'Procli Elementatio theologica translata a Guilelmo de Moerbeke', ed. H. Vansteenkiste, *Tijdschrift voor Philosophie*, 13 (1951), 263–302, 491–531.

Ps.-Aristoteles, 'Le Liber de Causis', ed. A. Pattin, *Tijdschrift voor Filosofi*, 28 (1966), 90–203.

Robert Grosseteste (Robertus Lincolniensis), *Commentarius in Posteriorum Analyticorum Libros*, ed. P. Rossi (Firenze: Leo Olschki, 1981).

Simon of Faversham, *Magistri Simonis Anglici sive de Faverisham Opera Omnia*, vol. 1, ed. P. Mazzarella (Padua: Cedam, 1957).

—, *Quaestiones super libro Elenchorum*, ed. by S. Ebbesen et al. (Toronto: Pontifical Institute of Medieval Studies, 1984).

Themistius, *In De Anima V—Commentaire sur le traite De l'ame d'Aristote, traduction de Guillaume de Moerbeke*, ed. G. Verbeke (Louvain: Publications universitaires de Louvain, 1957).

—, 'Themistius's Paraphrasis of the Posterior Analytics in Gerard of Cremona's Translation', ed. J. R. O'Donnell, *Medieval Studies*, 20 (1958), 239–315.

Studies

Christensen, M., 'Simon of Faversham *Quaestiones super De motu animalium*. A partial edition and doctrinal study', *Cahiers de l'Institut du Moyen Âge Grec et Latin*, 84 (2015), 93–128.

Ebbesen, S., 'Gerontobiologiens Grundproblemer', *Museum Tusculanum*, 40–43 (1980), 269–88.

—, 'Introduction', in Simon of Faversham, *Quaestiones super libro Elenchorum*, ed. by S. Ebbesen et al. (Toronto: Pontifical Institute of Medieval Studies, 1984).

—, 'Simon of Faversham. Quaestiones super librum De somno et vigilia. An Edition', *Cahiers de l'Institut du Moyen Âge Grec et Latin*, 82 (2013), 90–145.

Gibson, S., 'Confirmation of Oxford Chancellors in the Lincoln Episcopal Registers', *English Historical Review*, 26 (1911), 501–12.

Grabmann, M., *Die Aristoteleskommentare des Simon von Faversham: Handschriftliche Mitteilungen* (Munich: Verlag der Bayerischen Akademie der Wissenschaften, 1933).

Great Britain, Public Records Office, *Calendar of Close rolls. Edward I, vol. 5* [2?]: *1302–1307*, London.

Little, A. G., and F. Pelster, *Oxford Theology and Theologians ca. A.D. 1289–1302* (Oxford: Clarendon Press, 1934).

Longeway, J., *Simon of Faversham's Questions on the Posterior Analytics: A Thirteenth-Century View of Science*, Ph.D. in Philosophy (Ithaca: Cornell University, 1977).

Ottaviano, C., 'Le *Quaestiones super libro Praedicamentorum* di Simone di Faversham', *Memorie della Classe di Scienze Morali, Storiche e Filologiche*, 3 (1930), 257–351.

Pinborg, J., 'Simon of Faversham's *Sophisma: Universale est Intentio*: A Supplementary Note', *Medieval Studies*, 33 (1971), 360–65.

Register of John Pecham, Archbishop of Canterbury 1279–1292, ed. F. N. Davis et al., 2 vols. (Torquay: Devonshire Press, 1968–69).

Registrum Epistolarum Fratris Johannis Peckham, Archiepiscopi Cantuariensis, ed. C. T. Martin, 3 vols., Rerum Brittanicarum Medii Ævi Scriptores, or Chronicles and Memorials of Great Britain and Ireland during the Middle Ages, 77 (London: Longman & Co., 1882–55).

Registrum Ricardi de Swinfield, Episcopi Herefordensis, A.D. MCCLXXXIII–MCCCXVII, ed. W. W. Capes (London: Canterbury and York Society, 1909)

Registrum Roberti Winchelsey, Cantuariensis Archiepiscopi, A.D. 1294–1313, ed. R. Graham, 2 vols. (Oxford: Oxford University Press, 1956).

Salter, H. E., *Snappe's Formulary and Other Records* (Oxford: Clarendon Press, 1924).

Sharp, D., 'Simonis de Faverisham *Quaestiones super tertium De anima*', *Archives d'histoire doctrinale et littéraire du moyen âge*, 9 (1934), 307–68.

Shooner, H. V., *Codices Manuscripti operum Thomae de Aquino*, vol. 2 (Roma: Editori di San Tommaso, 1973).

Vennebusch, J., 'Die *Quaestiones in tres libros De anima* des Simon von Faversham', *Archiv für Geschichte der Philosophie*, 47 (1965), 20–39.

Weijers, O., and M. Calma, *Le travail intellectuel à la Faculté des arts de Paris: textes et maîtres (ca. 1200–1500) S–Z* (Turnhout: Brepols, 2012).

Wolf, F. A., *Die Intellektslehre des Simon von Faversham nach seinem De-anima-Kommentaren*, Inaugural-Dissertation (Bonn: Universität, 1966).

Yokoyama, T., 'Simon of Faversham's *Sophisma: Universale est intentio*', *Medieval Studies*, 31 (1969), 1–14.

Zimmermann, A., *Verzeichnis Ungedruckter Kommentare zur Metaphysik und Physik des Aristoteles aus der Zeit von etwa 1250–1350* (Leiden: Brill, 1971).

About the Team

Alessandra Tosi was the managing editor for this book.

Adèle Kreager proofread the book.

This book was refereed for the Medieval Text Consortium by Cecilia Trifogli and Rega Wood.

Jan Maliszewski typeset the book in LaTeX.

This volume uses the Tex Gyre Pagella font family.

Jeevanjot Kaur Nagpal designed the cover of this book.

John Longeway began this project decades ago, and we are very grateful for his allowing us to collaborate in bringing this volume at long last into print.

www.ingramcontent.com/pod-product-compliance
Lightning Source LLC
Chambersburg PA
CBHW050202240426
43671CB00013B/2218